NUREG/CR-7149
SAND2009-6323P

Effects of Degradation on the Severe Accident Consequences for a PWR Plant with a Reinforced Concrete Containment Vessel

Office of Nuclear Regulatory Research

AVAILABILITY OF REFERENCE MATERIALS
IN NRC PUBLICATIONS

NRC Reference Material

As of November 1999, you may electronically access NUREG-series publications and other NRC records at NRC's Public Electronic Reading Room at http://www.nrc.gov/reading-rm.html. Publicly released records include, to name a few, NUREG-series publications; *Federal Register* notices; applicant, licensee, and vendor documents and correspondence; NRC correspondence and internal memoranda; bulletins and information notices; inspection and investigative reports; licensee event reports; and Commission papers and their attachments.

NRC publications in the NUREG series, NRC regulations, and Title 10, "Energy," in the *Code of Federal Regulations* may also be purchased from one of these two sources.
1. The Superintendent of Documents
   U.S. Government Printing Office Mail Stop SSOP
   Washington, DC 20402–0001
   Internet: bookstore.gpo.gov
   Telephone: 202-512-1800
   Fax: 202-512-2250
2. The National Technical Information Service
   Springfield, VA 22161–0002
   www.ntis.gov
   1–800–553–6847 or, locally, 703–605–6000

A single copy of each NRC draft report for comment is available free, to the extent of supply, upon written request as follows:
Address: U.S. Nuclear Regulatory Commission
         Office of Administration
         Publications Branch
         Washington, DC 20555-0001
E-mail: DISTRIBUTION.RESOURCE@NRC.GOV
Facsimile: 301–415–2289

Some publications in the NUREG series that are posted at NRC's Web site address http://www.nrc.gov/reading-rm/doc-collections/nuregs are updated periodically and may differ from the last printed version. Although references to material found on a Web site bear the date the material was accessed, the material available on the date cited may subsequently be removed from the site.

Non-NRC Reference Material

Documents available from public and special technical libraries include all open literature items, such as books, journal articles, transactions, *Federal Register* notices, Federal and State legislation, and congressional reports. Such documents as theses, dissertations, foreign reports and translations, and non-NRC conference proceedings may be purchased from their sponsoring organization.

Copies of industry codes and standards used in a substantive manner in the NRC regulatory process are maintained at—
       The NRC Technical Library
       Two White Flint North
       11545 Rockville Pike
       Rockville, MD 20852–2738

These standards are available in the library for reference use by the public. Codes and standards are usually copyrighted and may be purchased from the originating organization or, if they are American National Standards, from—
       American National Standards Institute
       11 West 42nd Street
       New York, NY 10036–8002
       www.ansi.org
       212–642–4900

Legally binding regulatory requirements are stated only in laws; NRC regulations; licenses, including technical specifications; or orders, not in NUREG-series publications. The views expressed in contractor-prepared publications in this series are not necessarily those of the NRC.

The NUREG series comprises (1) technical and administrative reports and books prepared by the staff (NUREG–XXXX) or agency contractors (NUREG/CR–XXXX), (2) proceedings of conferences (NUREG/CP–XXXX), (3) reports resulting from international agreements (NUREG/IA–XXXX), (4) brochures (NUREG/BR–XXXX), and (5) compilations of legal decisions and orders of the Commission and Atomic and Safety Licensing Boards and of Directors' decisions under Section 2.206 of NRC's regulations (NUREG–0750).

United States Nuclear Regulatory Commission

Protecting People and the Environment

NUREG/CR-7149
SAND2009-6323P

Effects of Degradation on the Severe Accident Consequences for a PWR Plant with a Reinforced Concrete Containment Vessel

Manuscript Completed: July 2010
Date Published: June 2013

Prepared by:
J. P. Petti, D. A. Kalinich, J. Jun, K.C. Wagner

Sandia National Laboratories
Operated by Sandia Corporation for the
U.S. Department of Energy
Albuquerque, New Mexico 87185

Prepared for
Office of Nuclear Regulatory Research
U.S. Nuclear Regulatory Commission
Washington, DC 20555

Jose Pires, NRC Project Manager

NRC Job Code N6371

Office of Nuclear Regulatory Research

ABSTRACT

Various forms of degradation have been observed in the containment vessels of a number of operating nuclear power plants in the United States. Examples of degradation include corrosion of the steel shell or liner, corrosion of reinforcing bars and prestressing tendons, loss of prestressing, and corrosion of bellows. The containment serves as the ultimate barrier against the release of radioactive material into the environment. Because of this role, compromising the containment could increase the risk of a large release in the unlikely event of an accident. Previous work in this area has assessed the effects that degradation has on the pressure retaining capacity of the containment vessel through structural analysis that account for degradation. These analyses have provided useful information about the effects of the degradation on the structural capacity of the containment in both deterministic and probabilistic fashions. However, the appropriate metric to use in assessing containment degradation effects during a severe accident was determined to require additional study. The previous work with probabilistic descriptions of the containment capacity were obtained from the results of the structural analysis models, and used as input for the risk models (Probabilistic Risk Assessment analyses). The risk was formulated in terms of the large early release frequency (LERF). The relative LERF values were computed for various postulated cases of degradation. In that study, an instance of degradation was treated as a change in the plant's licensing basis and assessed with U.S. NRC Regulatory Guide 1.174, "An Approach for Using Probabilistic Risk Assessment in Risk-Informed Decisions on Plant-Specific Changes to the Licensing Basis." The Regulatory Guide provides the limits of the acceptable increases in LERF due to changes in the plant. Many of the cases of postulated degradation were those consisting of local corrosion in the liner or shell that produce leaks that do not contribute significantly to LERF, and in some cases, cause no change in LERF. Early releases due to small leaks were found to contribute to the small early release frequency (SERF). Since Regulatory Guide 1.174 does not provide guidance on the limits of SERF, additional deterministic analyses were performed in this study to assess the effects of degradation on the consequences using metrics other than LERF. This study was performed using the Sandia codes MELCOR and MACCS. These codes were used to simulate two different accident scenarios (long- and short-term station black-outs) and compute the resulting consequences for a PWR plant with a reinforced concrete containment. The structural analyses used in the previous probabilistic study were used to develop the containment behavior models for the accident simulations. Several different postulated cases of liner corrosion were considered to enable a comparison of the consequences.

ABSTRACT

FOREWORD

This is the latest report of a multiple year study to obtain insights on the degree of degradation a containment can have and still be left in service without repair and without significant impact on risk. The goal was to enhance understanding of containment safety margins for degraded conditions. The study examined effects of degradation in containments using deterministic and risk-informed methods.

Initially, the study evaluated the effects of degradation on several types of containments with respect to the guidelines given in Regulatory Guide 1.174, "An Approach for Using Probabilistic Risk Assessment in Risk-Informed Decisions on Plant-Specific Changes to the Licensing Basis." It integrated fragility curves, developed for non-degraded and postulated degraded conditions using structural analysis, with pre-existing probabilistic risk assessment models used in NUREG-1150, "Severe Accident Risks: An Assessment for Five U.S. Nuclear Power Plants." A conclusion from the first part of the study was that several cases of postulated degradation involving corrosion of the liner or shell showed small increases, no increases or even decreases in Large Early Release Frequency (LERF). This is because the postulated corrosion degradation leads to containment failure by leakage, with an increase in Small Early Release Frequency (SERF), rather than rupture.

Following that conclusion, NRC examined the effects of liner degradation using metrics other than LERF. Additional deterministic analyses were performed and documented here to assess the effects of liner degradation on consequences using metrics other than LERF. The report documents a method that integrates containment fragility functions for non-degraded and postulated degraded conditions with accident progression analyses and offsite consequence assessments made with the Sandia National Laboratories-developed codes MELCOR and MACCS.

The method reported is valuable to inform case-by-case examination of containment degradation effects. The methods and results in this report and in reports for the earlier and preliminary phases of this study (NUREG/CR-6920, "Risk-Informed Assessment of Degraded Containment Vessels," and NUREG/CR-6706, "Capacity of Concrete and Concrete Containment Vessels with Corrosion Damage") are valuable to inform regulatory decisions that relate to the performance of a containment structure in a degraded condition such as: license renewal reviews and inspection, monitoring and repair of potentially degraded containments. They are also valuable to support other studies or research programs by informing them of the relative risk significance of potential degradation processes, and to support updating of guidance that relates to the performance of degraded or non-degraded containments for beyond design basis severe accidents.

Michael J. Case, Director
Division of Engineering
Office of Nuclear Regulatory Research
U.S. Nuclear Regulatory Commission

TABLE OF CONTENTS

LIST OF FIGURES

LIST OF TABLES

ACKNOWLEDGEMENTS

The project team acknowledges the guidance and support from the NRC project managers over the life of the program, Jose Pires and Herman L. Graves III, Office of Research, U.S. Nuclear Regulatory Commission.

Special thanks are extended to everyone involved for their support and contribution to this project. The project team would also like to thank Sandia colleagues, Michael F. Hessheimer, Randall O. Gauntt, and Nathan E. Bixler, as well as former Sandia colleague Jason V. Zuffranieri

Sandia is a multiprogram laboratory operated by Sandia Corporation, a Lockheed Martin Company, for the United States Department of Energy's National Nuclear Security Administration under Contract DE-AC04-94AL85000.

1. INTRODUCTION

Various forms of degradation have been observed in the containment vessels of a number of operating nuclear power plants in the United States. Examples of degradation include corrosion of the steel shell or liner, corrosion of reinforcing bars and prestressing tendons, loss of prestressing, and corrosion of bellows. The containment serves as the final barrier between the reactor vessel and related components and the environment. Because of this critical role, compromising the containment could increase the risk of a release of radionuclides to the surrounding environment in the unlikely event of a severe accident.

A series of previous studies have investigated the ability of degraded containments to resist leakage under overpressurization. Cherry and Smith (2001) investigated how corrosion in the steel shell or liner influences the predicted failure pressures of steel, reinforced concrete, and prestressed concrete containments. Specific effects related to the loss of prestressing force on the pressure capacity of prestressed concrete containment vessels were studied by Smith (2001).

These studies found, in general, that the pressure capacities of containments were in excess of the design pressures, even with the presence of significant degradation. These computed failure pressures are critical in assessing the effects during a severe accident since those conditions can produce pressures far exceeding the design pressure. This general conclusion is supported by testing performed at Sandia National Laboratories over the past 25 years (Hessheimer, 2003). These tests proved to be of significant importance in assessing the behavior of different types of containment vessels and when performing analytical studies.

Even though the previous analytical studies provided valuable information on the pressure capacity of degraded containments, they did not study how degradation may affect the risk of a radioactive release. The initial step in the coupling of deterministic structural analyses of degraded containments within a risk framework was introduced by Ellingwood and Cherry (1999). They developed fragility curves for a steel containment with corrosion damage under overpressurization by performing multiple analyses with statistical variation in the material properties and other uncertainties. This method was used to compare the effects of degradation in a probabilistic setting.

Building on the work by Ellingwood and Cherry (1999), Spencer et al. (2006) integrated structural analysis results and their fragility curves with pre-existing probabilistic risk assessment (PRA) models to gain a risk-informed perspective on the issue of containment degradation. The PRA models used were originally developed for NUREG-1150, "Severe Accident Risks: An Assessment for Five U.S. Nuclear Power Plants," and NUREG/CR-4551 (Volume 1), "Evaluation of Severe Accident Risks: Methodology for the Containment, Source Term, Consequence and Risk Integration Analysis." The coupled structural-risk approach was applied to case studies of containment degradation at four "typical" U.S. nuclear power plants explored in NUREG-1150. This methodology enabled not only the determination of the risk, in terms of the large early release frequency (LERF), but also the changes in LERF due to postulated cases of degradation. The limits of the acceptable increases in LERF are provided in Regulatory Guide 1.174 (U.S. NRC, 2001) for changes in the plant's license basis. Here, the increase in LERF due to an instance of degradation was treated as such a change.

Typical degradation in the form of local corrosion in the liner or shell made up the majority of the postulated cases of degradation considered by Spencer et al. (2006). However, local corrosion in the liner of a typical reinforced or prestressed concrete containment was shown to only produce a noticeable effect on the pressure at which a leak would initiate, and not the pressure at which larger rupture or catastrophic rupture failures would occur. Due to the binning process used in the NUREG-1150 PRA models, degradation cases that only affect the leak pressure do not contribute to LERF. In some cases, LERF was shown to decrease due to earlier leaks caused by local corrosion. This was due to these earlier small leak failures in the PRA analysis being binned into the small early release frequency (SERF) instead of LERF. In the NUREG-1150 PRA models, leaks were defined as breaches that would not lead to depressurization in less than 2 hours. Therefore, outcomes in the PRA analysis leading to a leak failure of the containment were therefore included in SERF not LERF. Regulatory Guide 1.174 provides guidance for limits on changes in LERF and not on the limits for SERF. There is currently no appropriate guidance for assessing the risk significance of degradation in containment structures that cause only increases in SERF. Since LERF and SERF are only surrogates for the release consequences, the current study includes deterministic source term and consequence analyses to assess the effects of degradation on the consequences using metrics other than LERF.

The consequence analyses performed in this study employ two codes developed at Sandia National Laboratories, MELCOR and MACCS. MELCOR 1.8.6 (Gauntt et al., 2005) is a fully integrated, engineering-level computer code that models the progression of severe accidents in light water reactor nuclear power plants. MELCOR is being developed at Sandia National Laboratories for the U.S. Nuclear Regulatory Commission as a second-generation plant risk assessment tool and the successor to the Source Term Code Package. A broad spectrum of severe accident phenomena in both boiling and pressurized water reactors is treated in MELCOR in a unified framework. These include thermal-hydraulic response in the reactor coolant system, reactor cavity, containment, and confinement buildings; core heat-up, degradation, and relocation; core-concrete attack; hydrogen production, transport, and combustion; fission product release and transport behavior. Current uses of MELCOR include estimation of severe accident source terms and their sensitivities and uncertainties in a variety of applications. The MACCS code (Chanin, 1990) was developed for the U.S. Nuclear Regulatory Commission (NRC) to evaluate offsite consequences of hypothetical severe accidents at nuclear power plants (NPPs) based on calculated or assumed fission product releases. The code is designed to take input directly from the fission product releases calculated by MELCOR. MACCS also incorporates geographic, demographic, and meteorological data for a given plant, as well as assumptions concerning biological uptake. The output results can be expressed in terms of both health consequences and economical impacts. However, the present study only considers the health consequences in terms of dose, as well as prompt and latent cancer fatalities at different distances and times. As stated previously, these estimates are based on the MELCOR prediction for release and other representative assumptions for the local geography and meteorological conditions.

Since the analyses performed in MELCOR and specifically for the MACCS calculations are plant specific, an actual plant and accident scenario were chosen for this exploratory study. This was similar to the actual plants used in the NUREG-1150 study and the previous work on degradation effects by Spencer et al. (2006). Here, we examined a station blackout accident scenario at the Surry Power Station. Surry is a PWR plant with a subatmospheric reinforced concrete containment. The long- and short-term station blackout accident scenarios were selected since they produce the high internal pressures necessary to initiate cracks in the steel liner and outer concrete containment wall. In addition, the station blackout scenarios have been identified by NRC Standardized Probabilistic Analysis Risk (SPAR) models as being potentially important risk contributors. The station blackout scenarios are also commonly identified as important contributors in PRAs due to their common failure mode nature as well as the fact that both the containment and reactor safety systems are affected.

The degradation postulated for this containment was restricted to local corrosion in the steel liner near the wall-basemat junction and at the mid-height of the containment. The structural analyses of the reinforced concrete containment at Surry which incorporate the effects of these postulated cases of degradation were completed during the previous study by Spencer et al. (2006). Those results were used in this work to develop crack open area vs. internal pressure distributions for each of the postulated corrosion cases. These area vs. pressure curves are then introduced into the containment representation within the MELCOR analyses. The resulting source term output for each case of degradation are used as input for the MACCS code to determine the health consequences. The changes in consequences due to differences in corrosion location and extent are compared to the analyses of the containment without degradation. Using the Spencer study, three sets of analyses were performed: best estimate, lower bound, and upper bound. These bounding calculations are based on the crack area vs. pressure response space as determined in the structural analyses performed in the Spencer study. Even though the structural models, MELCOR, and MACCS models were used for the Surry Power Station, the cases of degradation assumed in this study are, as stated previously, hypothetical in nature, and do not reflect the actual conditions at Surry or at any other existing nuclear power plant. This study was conducted solely to demonstrate an analysis methodology and to examine the sensitivity of the consequences to varying cases of degradation.

In addition to the new deterministic source term and consequence analyses, this study includes ΔLERF results for several new cases of degradation. These new cases of degradation are modified versions of the cases examined in the previous study by Spencer et al. (2006). These cases are based on the same cases of degradation used previously, but include multiple instances of each simultaneously. The single cases of corrosion examined previously, were local in nature, extending only over several square inches. Since corrosion at an actual plant may be much more widespread, additional analyses which include more extensive degradation are also included in this study. The goal of these new PRA analyses, along with the deterministic MELCOR and MACCS analyses, is to further the state of knowledge regarding the appropriate metrics for assessing degradation in existing containment vessels. This can also provide valuable insights into the performance of the containment under beyond-design-basis loading events. An improved understanding of the significance and influence of the various factors that contribute to the risk and consequences may lead to important changes in the decision-making process when determining how to address these evolving safety issues at aging nuclear power plants.

2. CONTAINMENT PERFORMANCE CHARACTERIZATION

In order to perform a severe accident simulation, a representation of the containment performance under internal pressurization must be developed. When studying the pressure capacity of containment vessels, it is important to have an understanding of how these structures behave under this loading. The most critical, and difficult, task is to choose and then measure the performance criterion of a containment structure using the results of a structural analysis.

One of the potential measurements of a containment vessel undergoing pressurization is the leak rate at a given pressure. In addition, a given containment will also have a limiting pressure where the structure either catastrophically ruptures, or at a minimum, the containment contains sufficient leak paths to no longer be capable of remaining pressurized. These structural capacity limit pressures are also a critical performance characteristic, however, they are often of lesser importance since they are usually larger than the pressure at which the containment first begins to leak. The study by Spencer et al. (2006) showed that this was especially true for reinforced and prestressed concrete containment vessels.

In the PRA analyses performed for NUREG-1150, the containment performance was characterized by three categories of failures: "Leak", "Rupture", and "Catastrophic Rupture". Even though the leak rate for an actual containment vessel is a continuous function of the internal pressure, NUREG-1150 uses distinct categories for computational convenience in the binning process. A leak is defined as "… an opening that would arrest a gradual containment pressure buildup but would not result in containment depressurization within 2 h." "… A rupture is an opening that would result in rapid (<2 h) containment depressurization." A catastrophic rupture "… would eliminate major portions of the containment structures". These definitions are used in NUREG-1150 and in the previous study by Spencer et al. (2006) to estimate specific sizes of holes corresponding to the threshold leak rates for leak, rupture, and catastrophic rupture.

In the NUREG-1150 and the previous study by Spencer et al. (2006) the pressures indicating the onset of leak, rupture, and catastrophic rupture are cast in terms of fragility curves, or probabilistic representations of the failure pressure. In order to construct fragility curves from deterministic structural analyses, a series of analyses were required which included variations in the material and modeling uncertainties. For the current study, the same procedures used in the previous work by Spencer et al. (2006) are employed to compute the fragility curves for the PRA analyses in this study. However, several cases with more extensive degradation are examined in addition to those explored by Spencer et al. (2006). In addition, the source term and consequence analyses performed here with MELCOR and MACCS make use of the same structural analyses as those used for the PRA analyses. The MELCOR and MACCS analyses do not use fragility curves as input, but employ the leak area (total crack area) as a function of the internal pressure for a specific case. This section describes the general procedures used to model the containment capacity and compute the failure pressures and leak area vs. pressure distributions.

2.1 Containment Modeling

This study uses the analysis results from structural models developed under the previous study by Spencer et al. (2006). These models examined two concrete and two steel containments in order to estimate the pressures at which the various performance limits are met. The previous study focused on specific performance limits ("leak", "rupture", and "catastrophic rupture") required for input into the PRA models. Here, these performance limits are used in the additional PRA analyses performed in this study. In addition, the leak area, or crack open area, as a continuous function of pressure is also of interest for input into the MELCOR analyses.

Leaks in the containment boundary are caused by tears in containment liners and shells and initiate in regions of stress concentration (e.g. welds, local geometric details, or corrosion). These stress concentrations are generally several orders of magnitude smaller than the outer dimensions of the containment structure. Therefore, the structural modeling approach taken by Spencer et al. (2006), and adopted here, is to employ global finite element models of the containment structure to compute displacements and strains near the regions of anticipated stress concentrations. These displacements and strains are used as boundary conditions for detailed finite element models of local details or degraded regions.

The finite element models used for the structural analyses were completed with the ABAQUS (HKS, 2002) general-purpose nonlinear finite element analysis program.

The general procedures used to determine the containment performance for a reinforced concrete containment are summarized here. These procedures are the same as used by Spencer et al. (2006) with the relevant portions extended here for use in the deterministic MELCOR and MACCS analyses.

2.1.1 Performance Characterization for a Reinforced Concrete Containment

2.1.1.1 Steel Liner Leak Criterion for Corroded Regions

When examining the failure of the steel liner of the concrete containment considered in this study, it is assumed that the "leak" criterion is met when a tear initiates. At internal pressures below the level at which a tear first initiates, testing shows that there will probably exist a small amount of leakage, but it is assumed to be below the level of the "leak" threshold as defined in the PRA analyses and is assumed to not be consequence significant for the MELCOR and MACCS analyses.

Difficulties lie in determining where the first tear(s) is likely to occur due to the large number of local details that can potentially produce stress concentrations. In this regard, engineering judgment was used to select details and regions that exhibit the potential for the manifestation of high stress concentrations at lower pressures. Since actual containments contain a complex series of penetrations, welds, and other details, it is more than possible that some critical details have not been included in these computations. In addition, failures are likely to occur at local details not appearing on structural drawings (e.g. poor welds). The sampling of details and modeling methods considered in this study are assumed to be sufficient to provide a reasonable balance between the actual structural behavior and the practical matter of modeling that behavior.

Tears in the liner are predicted to initiate when the effective plastic strain, $\varepsilon_{p,eff}$, from the finite element analysis reaches a limiting value. The baseline limiting value of $\varepsilon_{p,eff}$ is the plastic strain found from a uniaxial tension test, denoted as ε_{fail}. Various conditions may be present that can cause a tear to initiate when $\varepsilon_{p,eff}$ is less than ε_{fail}. Spencer et al. (2006) adopted a modified version of the approach by Cherry and Smith (2001) to account for these effects. Equation 2.1 illustrates the adjustment procedure used to modify the effective plastic strains from the analysis to account for the multiaxial stress state, the effect of corrosion on the failure strain, and the model sophistication, or gauge length:

$$\bar{\varepsilon}_p = f_m \, f_c \, f_{c-u} f_g f_{FEM-u} \, \varepsilon_{p,eff} \qquad (2.1)$$

The factors in this equation are defined as follows,

- f_m accounts for the effects of multiaxial stress state,
- f_c accounts for the presence of corrosion,
- f_{c-u} accounts for the uncertainty in f_c,
- f_g is a gauge length factor,
- f_{FEM-u} denotes a finite element modeling uncertainty factor,
- $\bar{\varepsilon}_p$ denotes the factored plastic strain, and
- ε_{fail} denotes the uniaxial tensile strength.

The factored plastic strain, $\bar{\varepsilon}_p$ accounts for all of these effects. Failure occurs when $\bar{\varepsilon}_p$ is greater than ε_{fail}.

The multiaxial stress state factor, denoted as f_m was developed by Hancock and Mackenzie (1976):

$$f_m = \frac{1.0}{1.648 \, e^{-(\sigma_1+\sigma_2+\sigma_3)/2\sigma_{vm}}}, \qquad (2.2)$$

where σ_i are the principal stresses and σ_{vm} is the von Mises stress.

Experiments by Cherry and Smith (2001) found that a corroded steel plate could reasonably be modeled by reducing the thickness of the plate uniformly. However, the failure strain in the corroded areas only reached approximately 50% of the no corrosion failure strain value. This is due to unevenness and pits present on the corroded surface producing local stress and strain concentrations. Therefore, corroded shell or liner regions are modeled with uniform thinning in addition to the application of a corrosion factor, f_c equal to 2.0. In the PRA analyses the uncertainty of this effect is accounted for with the factor, f_{c-u}, which is introduced as a random parameter which is statistically varied when developing the fragility curves. The value of f_{c-u} follows a lognormal distribution with 1.0 as the median and 0.2 as the β factor. In all other areas of the model, f_c and f_{c-u} are set equal to 1.0. For the containment representations used in the best estimate MELCOR/MACCS analyses, f_{c-u} is set equal to 1.0, and to the same values used in the Spencer study for the upper/lower bound analyses.

Additional adjustments are included to account for the analysis sophistication, or gauge length, with the factor, f_g. This factor reflects the observation that strains tend to localize in areas of minor imperfections in the material. These localized strains can be much higher than the strains in the rest of the material a short distance from the localization area. The value assumed for this factor was based on work be Tang et al. (1995) who employed a gauge length factor of 4.0 in the derivation of strain magnification factors used to represent typical details in reinforced and

7

prestressed concrete containments. In addition, f_{FEM-u} reflects the uncertainty in the finite element model and the local strains computed with the meshes developed, boundary conditions, as well as other assumptions. Here, we use a lognormal distribution with 1.0 as the median and 0.30 and 0.35 as the β factors for 3D models and shell models, respectively, to represent f_{FEM-u}. Again, the best estimate value used for the MELCOR/MACCS containment representation is set to 1.0, and to the same values used in the Spencer study for the upper/lower bound analyses.

2.1.1.2 Penetration Leak Criterion

In reinforced concrete containments, a "leak" is assumed to initiate when tearing occurs in the steel liner. Here, the results of analytical models and experiments on typical discontinuities have been applied as reported in a study reported by Tang et al. (1995) and Castro et al. (1993). In these investigations, strain magnification factors were developed to approximate the effects of discontinuities typical of those found in the liners of reinforced and prestressed concrete containments.

The four locations identified as being susceptible to early tearing reside at large steam penetrations, at the junction of the wall and the basemat, at personnel or equipment hatches, and at the springline. Detailed finite element models, and in some cases, experimental models, were developed in the work by Tang et al (1995) and Castro et al. (1993) to determine the extent of the strain concentrations at those locations. These studies produced a series of strain magnification factors that enable the computation of the plastic strain in a discontinuity region using the global strain calculated with a simple axisymmetric model of the containment. Figure 2.1 illustrates the magnification factors for the four discontinuities as functions of the normalized global strain. The normalized global strain is defined as the appropriate global strain quantity from the axisymmetric analysis divided by the yield strain of the liner material.

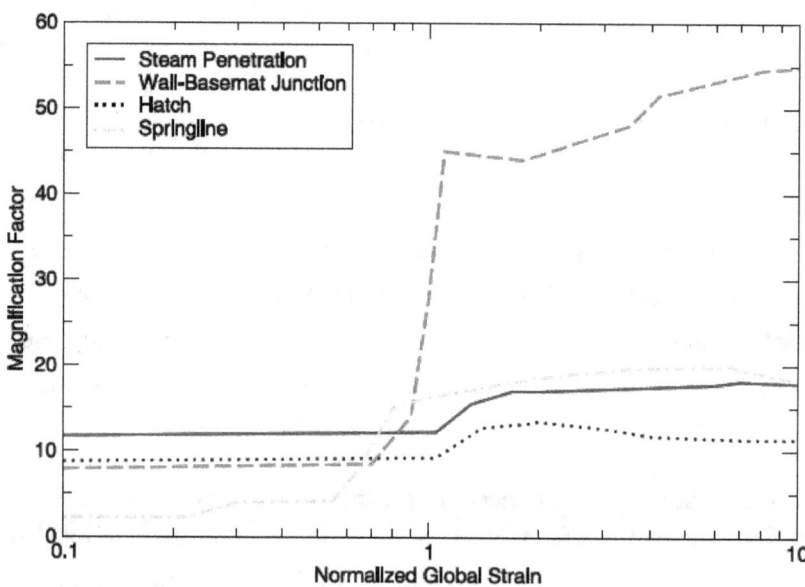

Figure 2.1: Strain Magnification Factors for Typical Discontinuities in Reinforced Concrete Containments (approximation of the curves from Tang et al., 1995)

As with the adjustment factors presented in the previous section for local analyses, these strain magnification factors adjust the global strain, ε_g, for multiaxial stress state, gauge length, and geometry effects.

The magnification factor, M, is determined using the appropriate curve in Figure 2.1. The magnified global strain, or the factored plastic strain, $\bar{\varepsilon}_p$, is then simply computed as:

$$\bar{\varepsilon}_p = M \; f_{FEM-u} \; \varepsilon_g \qquad (2.3)$$

where f_{FEM-u} denotes a finite element modeling uncertainty factor. Tearing failure is predicted to occur when the factored plastic strain, $\bar{\varepsilon}_p$, at the specific discontinuity location within an axisymmetric model at a given pressure exceeds ε_{fail}. In this study, corrosion was not assumed to exist at any of the typical discontinuities. Therefore, Eq. (2.3) does not include any adjustment for corrosion effects. For the MELCOR/MACCS analyses, f_{FEM-u} is equal to 1.0 for the best estimate analyses and to the same values used in the Spencer study for the upper/lower bound analyses.

2.1.1.3 Concrete Containment Rupture Criterion

In the NUREG-1150 PRA analyses, the hole size in the steel liner required to meet the rupture criterion is estimated between 0.028 m² (0.3 ft²) and 0.093 m² (1.0 ft²). Here we define rupture using the more conservative value of 0.028 m² (0.3 ft²). For the MELCOR/MACCS analyses, the distribution of the crack area vs. pressure as a continuous function is required. The analyses performed in the study by Spencer et al. (2006) did not explicitly model the actual crack (leak) initiation and growth, but used the tearing criterion summarized above. Therefore, an approximate method developed by Dameron et al. (1995) was employed to estimate the hole size. Given a location and orientation (meridional or hoop) of an assumed crack within a model, a crack width is calculated by

$$w = \varepsilon_g \; s \qquad (2.4)$$

where ε_g denotes the global strain and s equals the span or spacing between anchorages or stud layers perpendicular to the liner crack. The area of the opening can be estimated by multiplying the crack width by an assumed crack length, a.

$$A = w \; a \qquad (2.5)$$

The length of the crack can be estimated as the region of uniform peak strains or stresses, the size of local features, or the spacing of anchorages of studs in the direction of the crack. Equation 2.5 leads to an approximate estimate of the opening area since, among other factors, the opening is assumed square. Alternate estimates of the crack area can be constructed, but Equation 2.5 is assumed sufficient due to the large number of approximations employed here.

Since Equation 2.5 estimates the area for only one crack, the total open area at a given pressure is computed by

$$A_{total} = f_{rupt-u} \left(A_1 \delta_1 n_1 + A_2 \delta_2 n_2 + ... \right) \qquad (2.6)$$

where the subscripts 1 and 2 are associated with different locations within the model where cracks are assumed to have a possibility of initiating.

- A_i denotes the crack area for one crack at location i,

- δ_i equals 1 if the leakage criterion has been obtained at location i and equals 0 if leakage has not initiated,
- n_i denotes the number of cracks assumed to exist and initiate at location i, and
- f_{rupt-u} accounts for the uncertainty in the total area.

The variable, δ_i, enforces the requirement that a leak must initiate prior to a specific location contributing to the total open area. The crack number variable, n_i, enables the inclusion of the effects of multiple cracks initiating near the same location. Based on Equation 2.6, the total crack area is computed as a function of the internal pressure. For the PRA applications, a rupture is defined at the pressure when the total area, A_{total}, equals or exceeds 0.028 m^2 (0.3 ft^2). The MELCOR/MACCS analyses use the total area as a continuous function of internal pressure. Since the best estimate MELCOR/MACCS analyses use the median values for all properties, the area uncertainty, f_{rupt-u}, is set to 1.0. For the lower/upper bound analysis, the same uncertainties used in the Spencer study are employed.

2.1.1.4 Catastrophic Rupture Criterion

A simple method is used to estimate the onset of catastrophic rupture for use in the PRA analyses. By assuming a cylinder with contributions only from the rebar and liner, the pressure to cause catastrophic rupture is computed with

$$P_{cat-rupt} = f_{cat-u}(0.9)\left(\frac{\sigma_{rebar-ult}\ A_{rebar}}{R\ s_{rebar}} + \frac{\sigma_{liner-ult}\ t_{liner}}{R} \right) \tag{2.7}$$

where

- $\sigma_{rebar-ult}$ and $\sigma_{liner-ult}$, denote the ultimate strength for the rebar and liner,
- A_{rebar} is the area of the rebar,
- s_{rebar} denotes the spacing of the rebar,
- t_{liner} is the thickness of the liner,
- R equals the radius of the containment, and
- f_{cat-u} accounts for the uncertainty in the catastrophic rupture pressure.

In addition, Equation 2.7 includes a reduction factor of 0.9. This factor reflects overestimation of the simple method observed in the analysis and testing of a 1:4-scale prestressed containment. The 1:4-scale prestressed containment tested at Sandia National Laboratories (Hessheimer et al., 2003) failed at 1.42 MPa, while the simple ultimate strength estimated failure at 1.54 MPa (Dameron et al., 2000). The ratio of 1.42 MPa to 1.54 MPa is reduced slightly to 0.9.

10

3. DETERMINISTIC CONSEQUENCE ANALYSIS PROCEDURES

The consequence analyses performed in this study employ two codes developed at Sandia National Laboratories, MELCOR and MACCS, with descriptions below.

3.1 MELCOR

MELCOR 1.8.6 (Gauntt, et al., 2005) is a fully integrated, engineering-level computer code that models the progression of severe accidents in light water reactor nuclear power plants. MELCOR is being developed at Sandia National Laboratories for the U.S. Nuclear Regulatory Commission as a second-generation plant risk assessment tool and the successor to the Source Term Code Package. A broad spectrum of severe accident phenomena in both boiling and pressurized water reactors is treated in MELCOR in a unified framework. These include:

- Thermal-hydraulic response of the primary reactor coolant system, the reactor cavity, the containment, and the confinement buildings
- Core uncovering (loss of coolant), fuel heatup, cladding oxidation, fuel degradation (loss of rod geometry), and core material melting and relocation
- Heatup of reactor vessel lower head from relocated fuel materials and the thermal and mechanical loading and failure of the vessel lower head, and transfer of core materials to the reactor vessel cavity
- Core-concrete attack and ensuing aerosol generation
- In-vessel and ex-vessel hydrogen production, transport, and combustion
- Fission product release (aerosol and vapor), transport, and deposition
- Behavior of radioactive aerosols in the reactor containment building, including scrubbing in water pools, and aerosol mechanics in the containment atmosphere such as particle agglomeration and gravitational settling
- Impact of engineered safety features on thermal-hydraulic and radionuclide behavior
- The various code packages have been written using a carefully designed modular structure with well-defined interfaces between them. This allows the exchange of complete and consistent information among them so that all phenomena are explicitly coupled at every step. The structure also facilitates maintenance and upgrading of the code.

Initially, the MELCOR code was envisioned as being predominantly parametric with respect to modeling complicated physical processes (in the interest of quick code execution time and a general lack of understanding of reactor accident physics). However, over the years as phenomenological uncertainties have been reduced and user expectations and demands from MELCOR have increased, the models implemented into MELCOR have become increasingly best estimate in nature. The increased speed (and decreased cost) of modern computers (including PCs) has eased many of the perceived constraints on MELCOR code development. Today, most MELCOR models are mechanistic, with capabilities approaching those of the most detailed codes of a few years ago. The use of models that are strictly parametric is limited, in general, to areas of high phenomenological uncertainty where there is no consensus concerning an acceptable mechanistic approach.

Current uses of MELCOR often include uncertainty analyses and sensitivity studies. To facilitate these uses, many of the mechanistic models have been coded with optional adjustable parameters.

This does not affect the mechanistic nature of the modeling, but it does allow the analyst to easily address questions of how particular modeling parameters affect the course of a calculated transient. Parameters of this type, as well as such numerical parameters as convergence criteria and iteration limits, are coded in MELCOR as sensitivity coefficients, which may be modified through optional code input.

MELCOR modeling is general and flexible, making use of a "control volume" approach in describing the plant system. No specific nodalization of a system is forced on the user, which allows a choice of the degree of detail appropriate to the task at hand. Reactor-specific geometry is imposed only in modeling the reactor core. Even here, one basic model suffices for representing either a boiling water reactor (BWR) or a pressurized water reactor (PWR) core, and a wide range of levels of modeling detail is possible. For example, MELCOR has been successfully used to model East European reactor designs such as the Russian VVER, and RMBK-reactor classes. Current uses of MELCOR include estimation of severe accident source terms and their sensitivities and uncertainties in a variety of applications. The MELCOR Users Guide (Gauntt et al., 2005) provides extensive descriptions of the codes capabilities.

3.2 MACCS

The MACCS2 code (Chanin et al., 1990) was developed for the U.S. Nuclear Regulatory Commission (NRC) to evaluate offsite consequences of hypothetical severe accidents at nuclear power plants (NPPs) based on calculated or assumed fission product releases. The code is designed to take input directly from the fission product releases calculated by MELCOR. MACCS2 models the radioactive materials being released as being dispersed in the atmosphere while being transported by the prevailing wind. The specification of the release characteristics, often referred to as "source term", consists of Gaussian plumes (Systems Applications, 1982). Downwind population would be exposed to radiation, and land, buildings, and crops would be exposed to radioactive materials deposited from the plume. MACCS2 estimates the range and probability of the health effects induced by the radiation exposures and of the economic costs and losses that would result from the contamination of land, buildings, and crops. This section further describes a general calculation procedure of MACCS2. More detailed information on MACCS2 can also be found in the code manual for MACCS2 (Channin et al., 1990).

MACCS2 calculations are divided into modules and phases. This division is based on the sequence of societal response that would follow the occurrence of an accident. These phases are defined by the Environmental Protection Agency (EPA) in its *Protective Action Guides,* and referred to as the emergency, intermediate, and long-term phases (EPA, 1991). In addition, there are three distinctive modules within the MACCS2 calculation frame, and each module prepares for or performs calculations related to those three phases.

The first module, ATMOS, performs calculations related to atmospheric transport, dispersion and deposition, and the radioactive decay that occurs prior to release and while in atmosphere. Calculations in this module involve assessment of pre-accident inventories of radionuclides, transport and dispersion of radioactive material to offsite through development of a plume in the wake of plant buildings, and the subsequent downwind transport using a straight line Gaussian plume model. The results from the ATMOS module are then stored for use by other modules.

The second module, EARLY, performs calculations pertaining to the emergency phase which include user specified emergency response scenarios such as evacuation, sheltering, and relocation. Direct exposure pathways, dosimetry, and any mitigative actions and health effects during the emergency phase are evaluated in the EARLY module.

The third module, CRONC, models the events that occur following the emergency phase. The individual health effects that results from direct exposure to contaminated ground and from inhalation of resuspended materials, as well as indirect health effects caused by the consumption of contaminated food and water are calculated by the CRONC module. Additionally, CRONC simulates the economic consequences, such as the cost of the long-term protective actions or the cost of the emergency response actions that were modeled in the EARLY module.

In this study, the following input data were made available to perform the consequence analysis using MACCS2. Further discussions on the development of specific input parameters and assumptions used in the analysis are provided in Section 5.2.

- The radioactive nuclide inventory important for the consequence analysis at accident initiation.
- The atmospheric source term resulting from the accident including time and duration of release.
- Meteorological characteristics of the site region such as wind speed, atmospheric stability, and rainfall readings.
- The population distribution about the reactor site.
- Emergency response assumptions (evacuation time and average moving speed, effectiveness and occurrence of sheltering, criteria and timing for post-accident relocation of people, decontamination criteria and effectiveness, temporary interdiction criteria for land and buildings, and disposal criteria for contaminated crops).
- Land usage (habitable land fractions and farm land fractions) and economic data (crops worth, land, and buildings) for the region about the site.

In general, MACCS2 generates results for all possible combinations of a representative set of weather sequences and a representative set of exposed downwind population distributions, since consequences vary with source term magnitude, weather, and population density (Breeding et al., 1992). Hence, in MACCS2, results are reported in terms of statistical distributions of consequence measures such as doses, health effects, or costs. It should be noted, however, that the focus of this study is placed on relative health consequence results produced from different levels of containment degradation, not economic consequences.

4. PROBABILISTIC RISK ASSESSMENT PROCEDURES TO ASSESS CONTAINMENT DEGRADATION

The following section provides a summary of the PRA procedures used in the previous study by Spencer et al. (2006) to assess containment degradation. These same procedures have been adopted for the current study in examining several additional cases.

4.1 Guidelines for Assessing Containment Degradation Using Probabilistic Risk Assessments

The United States Nuclear Regulatory Commission (NRC) has recognized (NRC, 1995) the value of PRA techniques for nuclear applications in part due to improved safety and better allocation of limited resources, and has encouraged expanded use of PRA in regulatory decisions. Regulatory Guide 1.174 (USNRC, 2001) outlines the NRC's guidelines for using PRA methods when making changes to a nuclear power plant's licensing basis. Here, cases of degradation are treated as changes to the plant's licensing basis.

The five principles to guide a risk-informed licensing basis change as outlined in Regulatory Guide 1.174 are:
- "The proposed change meets the current regulations unless it is explicitly related to a requested exemption or rule change, i.e., a "specific exemption" under 10 CFR 50.12 or a "petition for rulemaking" under 10 CFR 2.802."
- "The proposed change is consistent with the defense-in-depth philosophy."
- "The proposed change maintains sufficient safety margins."
- "When proposed changes result in an increase in core damage frequency (CDF) or risk, the increases should be small and consistent with the intent of the Commission's Safety Goal Policy Statement. " (NRC, 1986)
- "The impact of the proposed change should be monitored using performance measurement strategies."

The fourth principle states that if a change results in an increase in risk, the increase in risk must be small and fall within certain acceptance guidelines. Regulatory Guide 1.174 designated both core damage frequency and large early release frequency (LERF) as criteria for assessing the proposed change. Since this study is only examining degradation in the containment vessel, only changes in LERF will be examined here.

Figure 4.1 illustrates the acceptance criteria for a proposed licensing basis change in terms of LERF. The change in LERF is plotted versus the baseline metric and includes three regions. Region I defines the conditions where the increase in the risk due to a proposed change is deemed too high to be acceptable, and the change is not permitted. For Region II, the risk increase is smaller than Region I, and the change is permitted but requires additional monitoring. In Region III, the change is permitted since the increase in the risk is defined as very small. In addition, the risk increases that fall in Region III are deemed so small enough that the value of the baseline risk metric does not need to be computed.

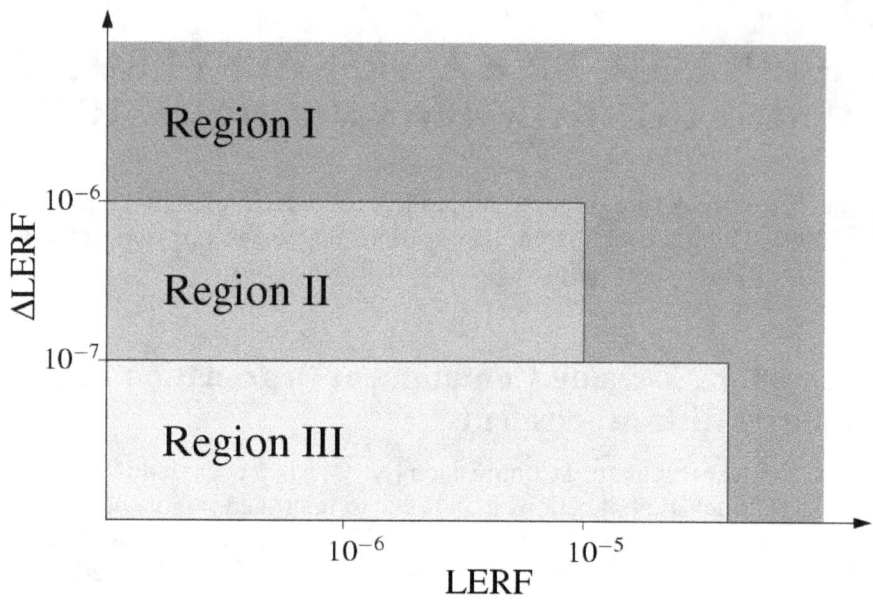

Figure 4.1: Acceptance Guidelines for Large Early Release Frequency

The application of Regulatory Guide 1.174 requires that the licensees clearly define the proposed change(s) and conduct sufficient analyses to demonstrate that the acceptance guidelines are met. The licensee must also present a methodology for implementing and monitoring the proposed change.

Since containment degradation reduces the pressure containing capacity of the structure, it is possible that determining the change in risk (e.g. LERF) may provide a means of assessing that degradation. In the study by Spencer et al. (2006), the guidelines of Regulatory Guide 1.174 were applied to assess the risk associated with various types of postulated degradation of containment vessels and bellows. The degradation was considered to be a temporary change in the plant's licensing basis using a coupled structural-PRA analysis and Regulatory Guide 1.174.

4.2   Procedure for Risk-Informed Containment Analysis

As applied in the study by Spencer et al. (2006) and in this study, the structural fragility under internal pressurization serves as the interface between the structural analyses and PRA models created for NUREG-1150. Since structural analyses are deterministic in nature, a probabilistic distribution of the structural capacity is required. This probabilistic distribution, or fragility curve, is the cumulative probability of failure for the containment under internal pressure. The criterion used to assess the potential structural failures was described in Section 2. In order to expand these deterministic, point values, a modified version of the methodology used by Ellingwood and Cherry (1999) was employed by Spencer et al. (2006). This method included uncertainty in the structural capacity using both the inherent randomness in the components that make up the structure and the uncertainty in the analyst's ability to characterize the behavior of the structure. These are often referred to as aleatory uncertainty and epistemic uncertainty, respectively.

The study by Spencer et al. (2006) includes a rigorous method to determine the structural fragility. This included estimating or determining the probability distributions associated with the individual components within the structure. Since the number of random variables required to perform a full Monte Carlo simulation is not reasonable for a complex containment structure, a Latin Hypercube Simulation (LHS) technique (McKay et al, 1979) was used to reduce the number of samples required. LHS analyses do not generate purely random variables for each sample; however, the stratum for each variable is randomly selected. The use of the LHS technique reduces considerably the number of structural analyses required to construct a structural fragility curve. Spencer et al. (2006) used the DAKOTA (Eldred et al, 2002) uncertainty analysis software to generate the LHS samples for the structural analyses.

In addition to the overall fragility curve, the NUREG-1150 PRA models also require the conditional probabilities of failure. The overall fragility describes the cumulative probability of failure as a function of the internal pressure, where failure is defined as any of leak, rupture, or catastrophic rupture. The conditional probability of failure describes the probability of each different type of failure and/or mode, given that there is a failure. Since the effects that different modes of failure have on the accident progression are critical, these differences must be accounted for during the compilation of the fragility curves.

For the given containment, thirty sets of random parameters were produced using the LHS procedure and used as input into the analytical models of the containment. Locations of potential failure are modeled using a combination of global and local finite element models. The failure criteria used to determine the pressure at which failure occurs at a specific location within the containment were described in Section 2.

A fragility curve at each location of potential failure is compiled using rank probabilities. For each of the 30 analyses, the failure pressures at a given location are ranked with each assigned a probability. The lowest pressure analysis is assigned the lowest failure probability equal to (1-0.5)/n, where n is 30 here. The next highest pressure is assigned a probability of (2-0.5)/n, or (i-0.5)/n for i=1,2…30. Fitted lognormal distributions are used to extrapolate the failure pressures to probabilities in the tail regions.

The fragility curves for all of the failures are combined into one cumulative probability of failure curve using de Morgan's rule (Ang and Tang, 1975). The probability of occurrence of any one of a number of statistically independent events E_n can be computed as

$$P(E_1 \cup E_2 \cup ... \cup E_n) = 1 - (1 - P(E_1))(1 - P(E_2))...(1 - P(E_n)). \qquad (4.1)$$

Failure in containments typically occurs at stress concentrations around penetrations, and the various penetrations in a containment can reasonably be assumed to be statistically independent. The penetrations modeled are just a sample of what may be a large number of very similar penetrations. The overall fragility is the cumulative distribution of the probability of a failure of any size. In addition, a larger size of failure at a location can only occur after the smallest size of failure (e.g. leak) at that location can occur; therefore, the distribution function for the smallest failure mode at every location is used in this calculation.

The conditional probabilities of failure of the three sizes (leak, rupture, and catastrophic rupture) are then computed. A leak is a breach that is smaller than the rupture size. A rupture is bigger than the leak size, but smaller than the catastrophic rupture size. The NUREG report by Spencer et al. (2006) describes the method used to compute the conditional probabilities. Using their method, the probability of each type of failure is given at a given pressure, given that a failure has occurred.

17

These cumulative and conditional failure probabilities are used as input into the PRA models developed in NUREG-1150 and described in detail in NUREG/CR-4551. In general, no detailed analyses were performed in that study to determine the containment performance/capacity. Instead, experts provided their estimates of failure pressure distributions using a number of methods including hand-calculations and experience. In addition, the original NUREG-1150 study did not investigate the effects of degradation in the containments. Since the purpose of the PRA analysis performed by Spencer et al. (2006) was to determine the change in LERF due to degradation, the PRA analyses are performed with the containment in its original and degraded states. Since containment degradation does not impact the core damage frequency, that portion of the original NUREG-1150 study is unchanged. The NUREG by Spencer et al. (2006) includes a short description of the PRA methodology and accident progression analyses. In the current study, we employ the same approach as Spencer et al. (2006), but examine additional cases of degradation.

While the study by Spencer et al. (2006) was successful in demonstrating a methodology to determine the change in LERF due to containment degradation, several cases of degradation proved to not affect LERF. In some cases, degradation even caused decreases in LERF. This could lead to the misleading conclusion that degradation may be beneficial from a risk standpoint. This is obviously not acceptable and shows that changes in LERF may not be the appropriate criteria to be applied in assessing containment degradation. Spencer et al. (2006) included the small early release frequency (SERF), as well as LERF. In the degradation cases that showed no change or decreases in LERF, SERF typically increased. Unfortunately, Regulatory Guide 1.174 does not provide guidance on acceptable changes in SERF.

Due to the results of the study by Spencer et al. (2006), the current work will examine additional cases of degradation to those considered previously. These additional cases will focus on including degradation extending over a larger area of the containment shell or liner. The previous analyses might have shown that LERF was relatively insensitive to degradation due to the regions of degradation being too small. These small regions of degradation, while causing earlier leaks, did little to affect the larger rupture and catastrophic rupture pressures. The binning process used in NUREG-1150 placed only these larger failures into LERF and the smaller leak failures into SERF. The more extensive degradation considered here is intended to produce larger effects on the rupture type failures, and consequently, LERF.

5. SEVERE ACCIDENT CONSEQUENCE ANALYSES OF PWR PLANT WITH REINFORCED CONCRETE CONTAINMENT

As examined by Spencer et al. (2006), a reinforced concrete (RC) containment for a pressurized water reactor (PWR) nuclear power plant was analyzed here to explore the effects of degradation during a severe accident. Corrosion in the steel liner wall was assumed near the wall-basemat juncture and at the mid-height of the containment as shown in Figure 5.1.

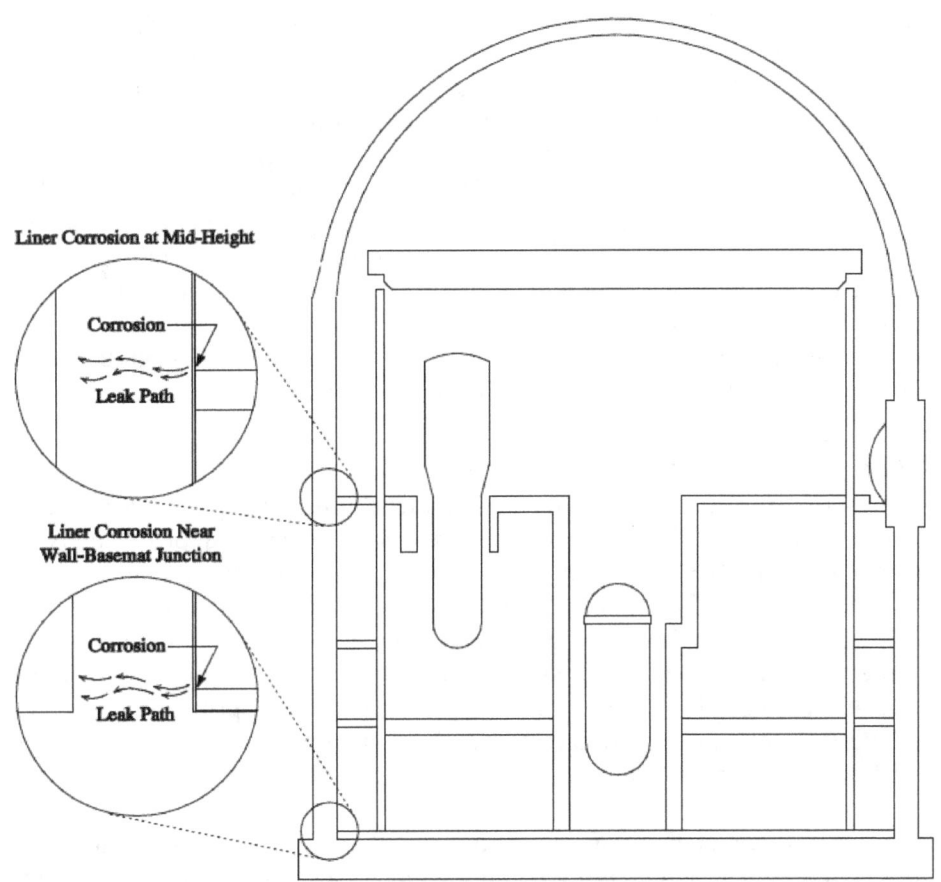

Figure 5.1: Postulated Corrosion in Liner of RC Containment (Spencer et al., 2006)

These locations were employed in this study and by Spencer et al. (2006) since these regions reside in areas of high liner strains and because water could accumulate in these regions. However, the study by Spencer et al. (2006) only explored analyses which included one small region of corrosion in the liner for each case. Here, we extended this work by assuming multiple locations of corrosion for a number of cases. These additional analyses were included to provide an understanding of the sensitivity of the risk and consequences to the extent of corrosion. The effects of the degradation on LERF are included in this study as well as deterministic analyses using MELCOR and MACCS. The structural analysis results from the study by Spencer et al. (2006) were used and modified in some cases for the present study.

Although the dimensions and properties of the Surry containment vessel were used to construct the structural model and site specifics data in the MACCS model, a number of assumptions have been made here including the hypothetical corrosion assumed for this study. Therefore, these analyses are not representative of actual conditions observed at that plant, or any other existing plant. This study was conducted to demonstrate an analysis methodology and to explore the potential effects that corrosion can have on risk and consequences.

5.1 Structural Model of Reinforced Concrete Containment

The experimental study by Horschel (1992) of a pressurization test of a 1:6 scale RC containment vessel with a steel liner showed that failures occur generally by tearing in the liner at geometric discontinuities. These tears then provide a leak path for material through the containment wall. Therefore, Spencer et al. (2006) assumed that a tear in the linear produced a failure of the containment. In addition, these tears would occur at penetrations or regions of corrosion. The modeling procedure used by Spencer is similar to that used by Cherry and Smith (2000). Cherry and Smith developed an axisymmetric model of the RC containment and determined the global strains in the liner. These global strains defined the boundary conditions for local models that include thinned areas caused by corrosion. A strain criterion was used in both studies to compute when tearing initiated. However, Spencer et al. (2006) include distinctions between leak, rupture, and catastrophic rupture to use in the PRA analyses, where Cherry and Smith only considered leak in their failure estimates. Here, we use the structural analysis results from the study by Spencer et al. for use in additional PRA analyses and in the MELCOR/MACCS analyses.

5.1.1 Axisymmetric Global Model

The axisymmetric finite element mesh of the containment shown in Figure 5.2(a) was used in the study by Spencer et al. (2006). The reinforced concrete containment is approximately 1.37 m (4.5 feet) thick with a 9.53 mm (0.375 inch) thick steel liner. The concrete is modeled using the ANACAP-U (ANATECH, 1997) constitutive model, which is linked with the ABAQUS (HKS, 2002) finite element program. Reinforcing bars are modeled by embedding them within the concrete elements. The liner is modeled with axisymmetric shell elements directly attached to the concrete elements on the inside face of the containment. Nonlinear spring elements are used to represent the interface conditions between the containment and the soil. These springs exhibit high elastic stiffness in compression with no strength in tension.

For the analyses by Spencer et al. (2006), the containment model was subjected to a monotonically increasing internal pressure loading with an increasing temperature. The temperature load is increased as the pressure loading increases. The initial temperature of the liner was set at 21° Celsius (70° Fahrenheit) and increases to a peak temperature of 176° Celsius (349° Fahrenheit) at a pressure of 1.38 MPa (200 psi).

Figure 5.2(b) shows the deformed shape (magnified by 10) of the axisymmetric model for a typical case near the end of the analysis.

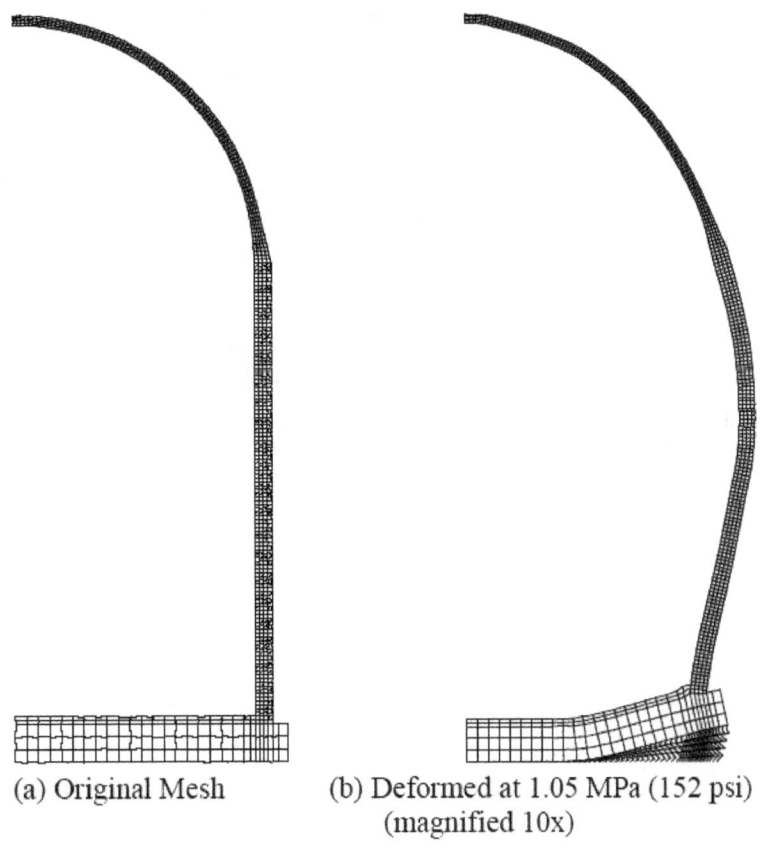

(a) Original Mesh	(b) Deformed at 1.05 MPa (152 psi)
	(magnified 10x)

Figure 5.2: Axisymmetric Finite Element Model of RC Containment (Spencer et al., 2006)

5.1.2 Local Detail and Corrosion Modeling

In Section 2.1.1.2, the strain magnification factors are described for use when approximating the strain concentrations in the liner at discontinuities. Section 2.1.1.1 provides the method used when assessing the detailed 2-D finite element models used to model regions of hypothetical corrosion. Two locations of liner corrosion were studied by Spencer et al. (2006), one at the mid-height of the containment and the other in the containment wall just above the junction with the basemat. Corrosion was assumed to have thinned the original liner at two different levels, 50% and 65% for each location. These are the percentages of material that is no longer in its near original condition. A thin layer of severely reduced strength corroded steel constituents would be attached, but is ignored using this uniform thinning method. Corrosion was assumed to have occurred over an in-plane region 8 cm (3 in) wide and 13 cm (5 in) high for the mid-height case as shown in Figure 5.3. The outer boundaries of the mesh cover a 91 cm (36 in) wide square region. Figure 5.4 shows the finite element mesh used in the basemat region of the containment. The corroded region is approximately 25 cm (10 in) wide and 18 cm (7 in) high. At the boundaries of the mesh, the average length of an element edge is 25 mm (1 in), while in the corroded area, the mesh is refined considerably and the average element length is approximately 8 mm (0.3 in). In both of these cases, a single layer of elements surround the corroded region to provide a gradual transition from the corrosion zone to the undamaged liner. In that layer of elements, the corrosion is assumed to penetrate 25% and 32.5% of the wall thickness for the 50% and 65% corrosion cases, respectively. Nonlinear spring elements are used to approximate the behavior of the shear studs that attach the liner to the concrete wall. The stud locations are

21

illustrated in the finite element meshes, and are typical of details that would be seen in an actual RC containment.

The size and shape of the corroded regions employed here are hypothetical in nature and not based on any specific corrosion data at any existing plant. They were developed in order to provide a large enough area of corrosion to manifest a sufficient stress/strain concentration where a crack would initiate. Actual corrosion in plants will vary widely in shape, extent, and level. This study's goal is to demonstrate a methodology and not to assess any specific plant. If specific plant corrosion data is available, a site specific analysis should be performed.

The results from the local models were post-processed to determine the point at which tearing occurs. For a tear to initiate, the effective plastic strains are multiplied by factors to account for the effects of the multiaxial stress state, corrosion, gauge length, and modeling uncertainty, as described in Section 2.1.1.1.

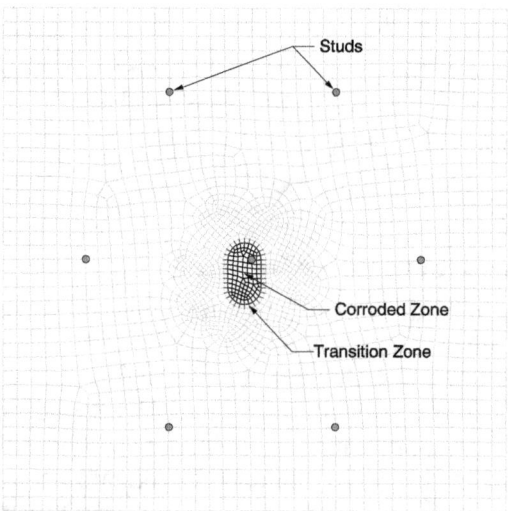

Figure 5.3: Finite Element Mesh of Corrosion at Midheight

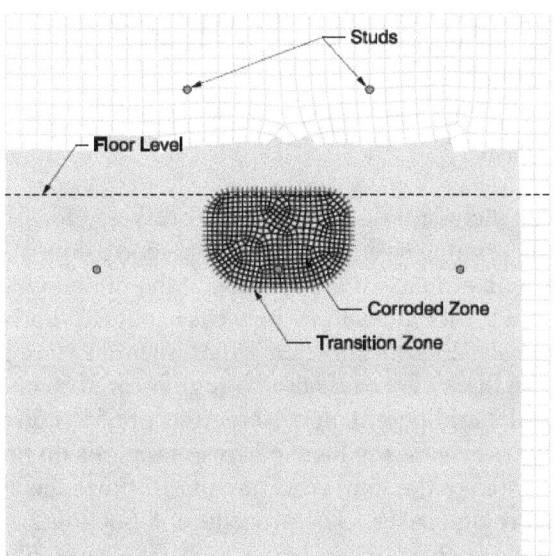

Figure 5.4: Finite Element Mesh of Corrosion in Wall Near Basemat

5.1.3   Structural Analysis with Best Estimate Properties

The focus of the structural analysis performed by Spencer et al. (2006) was to develop fragility curves for use as the interface with the NUREG-1150 PRA models. Their study also included a description of the baseline analysis results that used the median values of the input parameters. This baseline analysis was used here for input into the MELCOR accident analysis model, and therefore, is summarized below.

Six potential liner failure modes are considered: failure at the four locations of geometric discontinuities for which magnification factors are used (see Section 2.1.1.2), failure at a corroded region at the wall midheight, and failure at a corroded region on the wall near the basemat. To investigate the potential for failure (leak) at all of these locations, the global axisymmetric model is analyzed, and results are extracted for use as displacement boundary conditions for the two local corrosion models. The two local corrosion models are run with 50% and 65% corrosion and post-processed to find the most critical value of $\bar{\varepsilon}_p$ at every loading step using the leak criterion (see Section 2.1.1.1). A separate post-processing tool is applied to the results of the axisymmetric model to apply appropriate magnification factors to find $\bar{\varepsilon}_p$ for the four selected discontinuity locations at every loading increment. The uniaxial plastic failure strain, ε_{fail}, was assumed to be 25% for these analyses. The gauge length factor was assumed to be 4.0 for the local corrosion models in this baseline analysis.

Figure 5.5 shows the plot of the ratio of $\bar{\varepsilon}_p$ to ε_{fail} (0.25) for the 6 failure modes as a function of internal pressure for the baseline analysis. A horizontal line has been drawn at 1.0 to show the failure criterion. It is evident that the failure condition is met first at the corroded area at the midheight, and that the corroded area near the basemat appears to be the least critical location. Of the geometric discontinuities, the steam penetration is the most critical. From this analysis, it appears that a containment without any degradation would fail at 0.896 MPa (129.9 psi) at the steam penetration. If the postulated corrosion were present at the mid-height, it would fail at that

23

location at an internal pressure of 0.829 MPa (120.3 psi) and 0.812 MPa (117.8 psi) for 50% and 65% corrosion, respectively. Leak near the corroded basemat does not occur until 0.947 MPa (137.4 psi) and 0.939 MPa (136.2 psi) for 50% and 65% corrosion, respectively. Leakage occurs at the steam penetration prior to leak in the corroded basemat location. Table 5.1 summarizes the leak pressures for each location.

Figure 5.6 and Table 5.1 also include failure curves and pressures for five rupture cases examined by Spencer et al. (2006): the original containment with no corrosion, 50% corrosion at the basemat, 65% corrosion at the basemat, 50% corrosion at the midheight, and 65% corrosion at the midheight. In addition, Table 5.1 includes three new cases not explicitly examined under the previous study by Spencer et al. (2006). These cases assume that 10 regions of local corrosion exist in the containment. The three new cases are 10 regions of 50% corrosion at the basemat, 10 regions of 50% corrosion at the mid-height, and 10 regions of 65% corrosion at the mid-height. It is assumed that for each of these cases, the local corrosion regions do not interact. Therefore, the contribution of the corroded area to the total crack area is 10 times the amount that is used in the original, or single, corrosion region cases. The procedure in Section 2.1.1.3 is used to compute the ratio of the total rupture area, A_{total}, to the rupture area limit of 0.028 m^2 (0.3 ft^2), as a function of pressure. The four features, steam penetration, wall-basemat junction, hatch, and springline, are assumed to contain four cracks (n_i) each after the leak criterion has been reached.

The choice of four cracks at each location is based on the upper limit of the number of tears observed in the 1:6 scale reinforced concrete containment testing at Sandia National Laboratories (Horschel, 1992). The number of cracks at each location could have been introduced as a random variable for the PRA fragility analysis, but due to the severely approximate nature of the rupture estimate employed here, a random variable with a large uncertainty is imposed on the total rupture area. One crack is assumed at the basemat and midheight locations for cases considering corrosion. For each of the 10 regions of corrosion cases, 10 cracks are assumed.

For the two single basemat corroded cases, leak does initiate prior to the overall rupture failure. This has only a negligible effect on the total rupture area, since only one crack is assumed to initiate at the corroded location. Therefore, the rupture curves for the original case with no corrosion and the two basemat corrosion cases essentially share the same curve. A small increase in the rupture area is caused by a single region of corrosion at the midheight. Since the global strains do not differ between the 50% and 65% corrosion cases, the rupture curves are nearly identical. Obviously, the assumption of more than one corroded location for each case causes an increase in the rupture area and lowers the rupture pressure. This is especially true for corrosion at the midheight, where leak initiates at the corroded location well before initiation at the steam penetration. The effects of multiple basemat corroded locations are minimal since leak at the corroded basemat initiates just prior to the current rupture thresholds. Variation in the total area assumed to cause rupture (0.028 m^2 or 0.3 ft^2) will also affect the rupture pressure. As mentioned previously, a large uncertainty is imposed on the total rupture area to account for the approximate nature of the rupture criterion when computing the fragility curves in the next section. Finally, Figure 5.6 and Table 5.1 include the simple estimate for catastrophic rupture described in Section 2.1.1.4.

The next section describes the procedures used for developing the fragility curves used in the PRA analyses. For the deterministic MELCOR/MACCS analyses, three sets of analyses were performed. These include a best estimate (or nominal case), and lower and upper bound sets of analyses. The best estimate cases use the structural analysis results described in this section, where the lower and upper bound analyses use the structural analysis results from the analyses described in the next section on fragility curve development. Section 5.1.5 then provides the area vs. pressure curves for each set of analyses as they were extracted from the structural analyses.

Table 5.1: Summary of Locations and Failure Pressures

Location	Failure Pressure MPa (psig)
Leak Steam Penetration	0.896 (129.9)
Leak Wall-Basemat Junction	0.988 (143.3)
Leak Hatch	0.935 (135.6)
Leak Springline	0.992 (143.9)
Leak Base 50% Corrosion	0.947 (137.4)
Leak Base 65% Corrosion	0.939 (136.2)
Leak Midheight 50% Corrosion	0.829 (120.3)
Leak Midheight 65% Corrosion	0.812 (117.8)
Rupture No Corrosion	0.983 (142.6)
Rupture Base 50% Corrosion	0.983 (142.6)
Rupture Base 65% Corrosion	0.983 (142.6)
Rupture Midheight 50% Corrosion	0.976 (141.6)
Rupture Midheight 65% Corrosion	0.976 (141.6)
Catastrophic Rupture	1.195 (173.3)
Rupture Base 50% Corrosion 10x	0.975 (141.5)
Rupture Midheight 50% Corrosion 10x	0.920 (133.5)
Rupture Midheight 65% Corrosion 10x	0.920 (133.5)

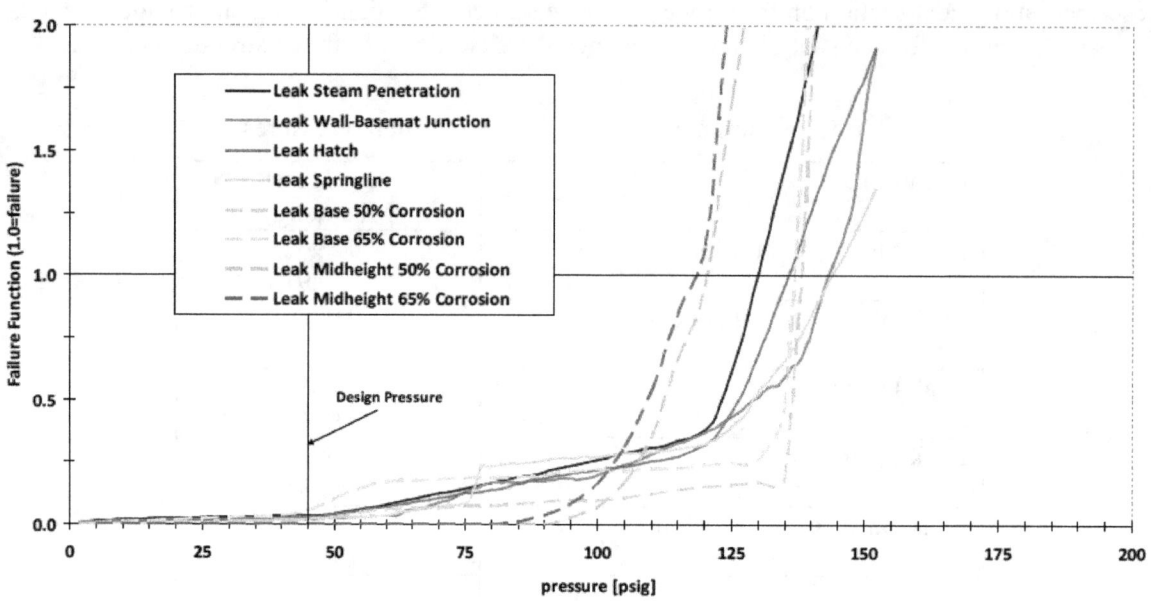

Figure 5.5: Failure Potential at Corrosion and Discontinuity Locations (Leaks)

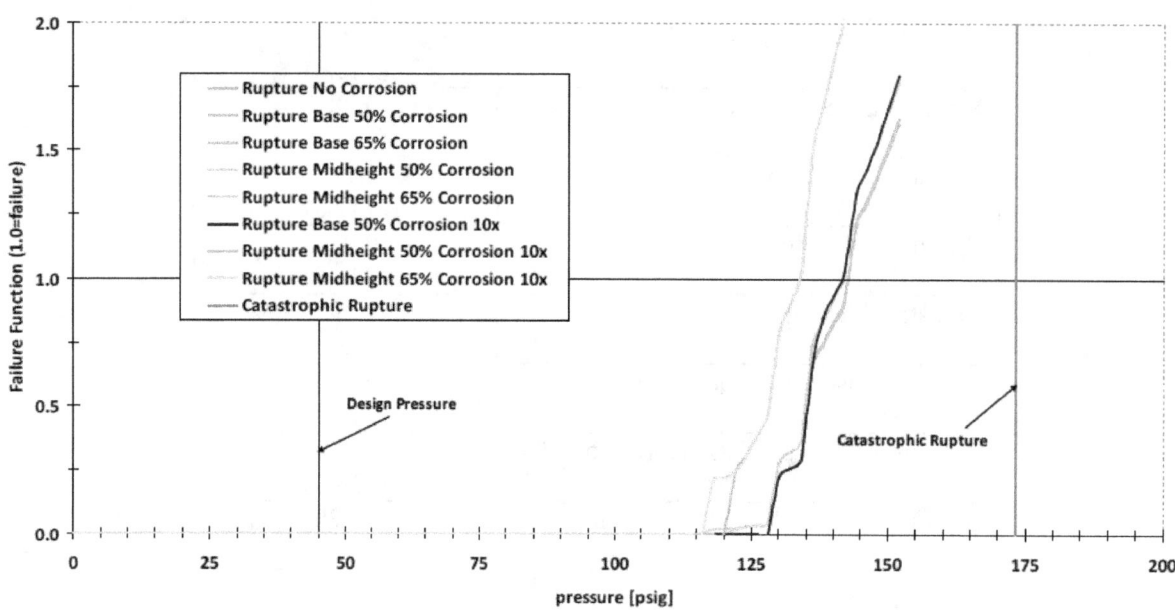

Figure 5.6: Failure Potential at Corrosion and Discontinuity Locations (Ruptures)

5.1.4 Fragility Curve Development for PRA Analyses

The principal goal of the study by Spencer et al. (2006) was to integrate structural analyses of degraded containments with risk analysis models to understand the significance of the degradation from a risk-informed perspective. Fragility curves of containments under pressurization with and without degradation serve as the interface between the structural and risk analyses. As explained in Section 4.2, Latin Hypercube Sampling was used to generate sets of random variables to be used as input to the structural finite element analyses. In the Spencer study, 19 input parameters were selected as random variables, and 30 samples of those parameters were developed for analysis.

Of the 19 input parameters, 14 describe the inherent randomness, or aleatory uncertainty, in the basic properties of concrete, reinforcing bars, and liner steel. The remaining 5 parameters describe epistemic uncertainty, or uncertainty in the ability to characterize the failure pressure. Several of the random input parameters are correlated with others.   For concrete, the tensile strength, f'_t, and modulus of elasticity, E_c, are correlated with the cylinder compressive strength, f'_c. Likewise, there is correlation between the yield strength and ultimate strength of steel.

Whenever possible, statistical test data was used as the basis for the probability distributions in this study. When such data was unavailable, engineering judgment was used to provide reasonable estimates of uncertainty. A thorough statistical analysis of concrete strength was performed by Mirza et. al. (1979).   Based on recommendations from that analysis, the elastic modulus of concrete is computed as:

$$E_c = c_1 60400\sqrt{f'_c} \tag{5.2}$$

where c_1 is a random variable with a median of 1.0 and a coefficient of variation of 0.077. Likewise, f'_t is computed as:

$$f'_t = c_2 8.3\sqrt{f'_c} \tag{5.3}$$

where c_2 has a median of 1.0 and a coefficient of variation of 0.196. Units of psi are used in both of these equations. Under average control conditions, Mirza et al. (1979) suggest that a coefficient of variation (COV) of 0.15 can be assumed for f'_c.

The RC containment investigated in this study has reinforcing bars of both Grade 40 and Grade 50 steel. Unfortunately, less information was available on the correlation between yield strength and ultimate strength, and between the elastic modulus and strength for steel. The elastic modulus, E_s, is assumed to be independent of the yield strength, f_y, and the ultimate strength, f_u. Correlation is induced between f_y and f_u in a manner similar to that for concrete:

$$f_y = s_1 r f_u \tag{5.4}$$

where s_1 is a random variable with a median of 1.0. The variable r is the ratio of the yield stress to the ultimate stress, and differs depending on the type of steel. For Grade 40 reinforcement, a value of 0.6 was assumed for r. For Grade 50 reinforcing, this value was set at 0.62, and for the liner steel, a value of 0.535 was used.

Five separate variables are used to describe the epistemic uncertainty. Variable describes the uncertainty inherent in predicting the global strains, the local strains from the 2-D submodels with

27

corrosion, the strain magnification in the presence of corrosion, the calculation of the rupture area, and the simple estimate for catastrophic rupture.

Each of these variables has a median value of 1.0. A COV of 0.3 was assumed for the global strain prediction variable, f_{FEM-u}. This agrees well with the scatter in the analytical predictions submitted by several analysts for the behavior of the 1:6 scale RC containment test conducted at Sandia National Laboratories (Clauss, 1987). The magnification factors for the liner discontinuities are simply multiplied by this uncertainty factor in Equation 2.3.

The COV of the local model uncertainty factor, f_{FEM-u}, was set to 0.35 because it was felt that there is more uncertainty in the local models than in the global models. For the local detailed models, Equation 2.1 is used to compute the magnified strains. Since the local models contain corrosion, the corrosion factor, f_c, equals 2 and is multiplied by an uncertainty factor, f_{c-u}, with a COV of 0.20.

The final two epistemic factors account for the uncertainty in the rupture area, f_{rupt-u}, and catastrophic rupture pressure, f_{cat-u}. Since the calculation of the rupture area in Equation 2.6 contains several approximations, a large COV is employed, equal to 0.40. The simple method used here to compute the catastrophic rupture pressure (Equation 2.7) should provide a reasonable estimate, enabling the use of a low COV of 0.10. This uncertainty is in addition to the variation in the ultimate strengths random variables used in Equation 2.7. Since different criteria are used for the rupture and catastrophic rupture failures, there is no correlation between their uncertainty factors.

Table 5.2 shows all of the random input parameters. This table shows the median values of the parameters, a measure of the variation assumed for the parameter, and the type of distribution. For variables with normal distributions, the COV is used as the measure of variation, while β is used for variables represented with a lognormal distribution.

A total of 30 sets of these input parameters were generated based on analyses using the specified distribution functions. Once the analyses were run, the 30 leak pressures were used to develop fragility curves for each of the four locations of linear discontinuity. Figure 5.7 shows the fragility curves based on the sorted analyses values and associated lognormal curve fits. The fragility curves represent the leak pressures for the four discontinuity locations. Figure 5.8 shows the fragility curves for leaks at the two corroded locations, basemat and midheight, at the two levels of corrosion, 50% and 65%. These plots indicate that increasing the level of corrosion from 50% to 65% of the liner thickness has only a minimal effect. Due to the shape of the distributions, the lognormal curve fits for each of the corroded cases require one curve fit for high probabilities and one curve fit for low probabilities.

Table 5.2: Random Input Parameters

Property	Median	COV/β	Distribution
Concrete			
Compressive Strength (f'_c)	27.3 MPa (3.956 ksi)	0.15	Lognormal
Elastic Modulus factor (c_1)	1.0	0.077	Normal
Tensile Strength factor (c_2)	1.0	0.196	Normal
Grade 40 Reinforcing Steel			
Ultimate Strength (f_u)	508.3 MPa (73.72 ksi)	0.09	Lognormal
Yield Strength Factor (s_1)	1.0	0.07	Normal
Elastic Modulus (E_s)	200 Gpa (29000 ksi)	0.06	Normal
Grade 50 Reinforcing Steel			
Ultimate Strength (f_u)	590.5 MPa (85.65 ksi)	0.09	Lognormal
Yield Strength Factor (s_1)	1.0	0.07	Normal
Elastic Modulus (E_s)	200 Gpa (29000 ksi)	0.06	Normal
Liner Steel			
Ultimate Strength (f_u)	577.1 MPa (83.70 ksi)	0.09	Lognormal
Yield Strength Factor (s_1)	1.0	0.07	Normal
Elastic Modulus (E_s)	200 GPa (29000 ksi)	0.06	Normal
Uniaxial Failure Strain (ε_{fail})	0.25	0.12	Normal
Corrosion Depth Uncertainty	1.0	0.10	Normal
Epistemic Uncertainty			
Axisymmetric Model Strain Factor Uncertainty (f_{FEM-u})	1.0	0.30	Lognormal
Submodel Model Strain Factor Uncertainty (f_{FEM-u})	1.0	0.35	Lognormal
Corrosion Factor Uncertainty (f_{c-u})	1.0	0.20	Lognormal
Leak Area for Rupture Uncertainty	1.0	0.40	Lognormal
Catastrophic Rupture Uncertainty	1.0	0.10	Lognormal

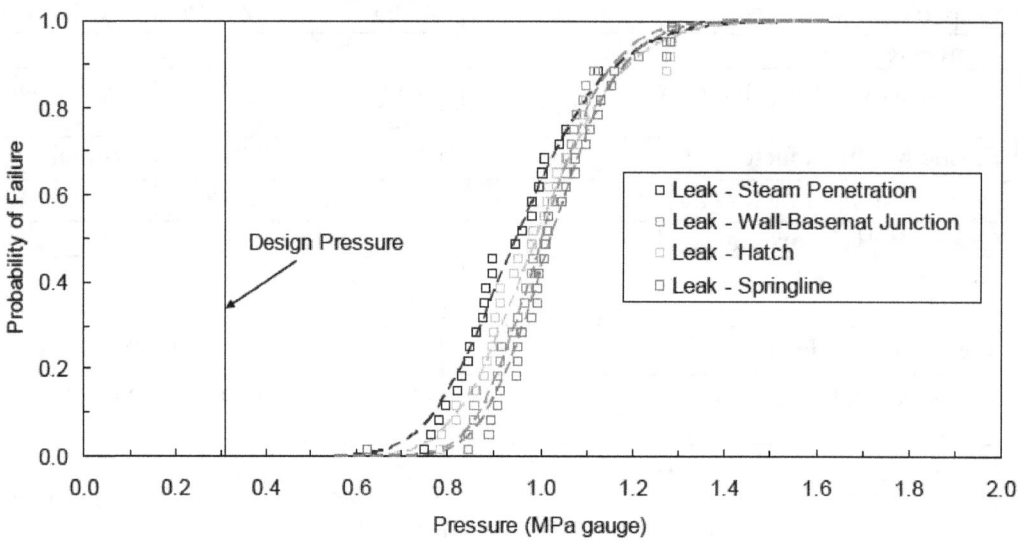

Figure 5.7: Fragility Curves for Leak at Discontinuity Locations

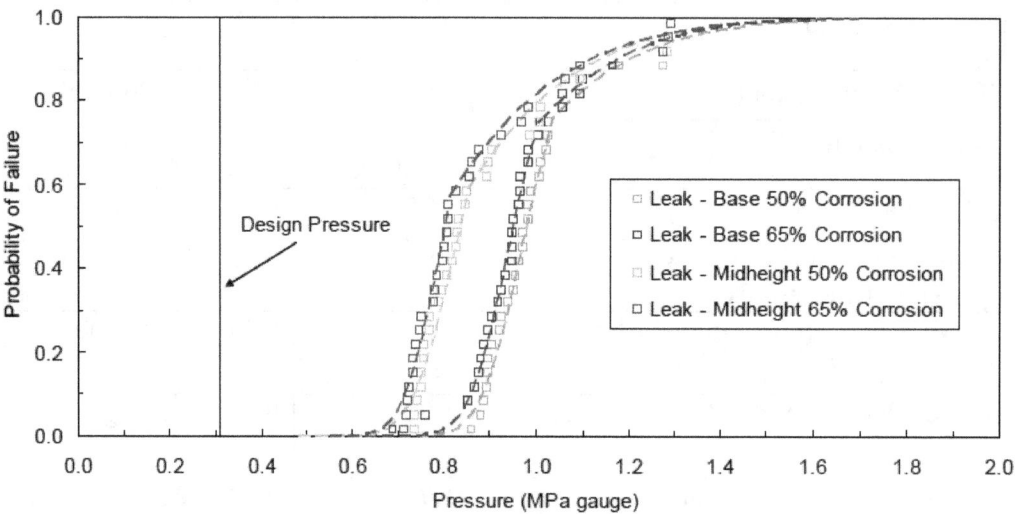

Figure 5.8: Fragility Curves for Leak at Corroded Locations

As discussed previously, the rupture threshold is assumed to be reached when the total crack opening caused by initiated leaks exceeds 0.028 m² (0.3 ft²). Figure 5.9 shows the rupture fragility curves for the eight cases examined in this study. The crack opening near the midheight corroded locations are large enough to cause a decrease in rupture pressure. Since the global strains are employed to estimate the rupture pressure, the single location 50% and 65% midheight corroded cases are the same. The 10 regions of corrosion cases at the midheight show a noticeable decrease in the rupture fragility, with the 65% decreasing slightly more than the 50% case. The basemat cases with a single corrosion location do not vary from the no corrosion case, while the 10 region case is only slightly decreased from the single corrosion cases. Figure 5.9 also illustrates the fragility curve for catastrophic rupture. Since the criteria for catastrophic rupture is not affected by the corrosion in the liner, the catastrophic rupture fragility remains the same for all eight cases.

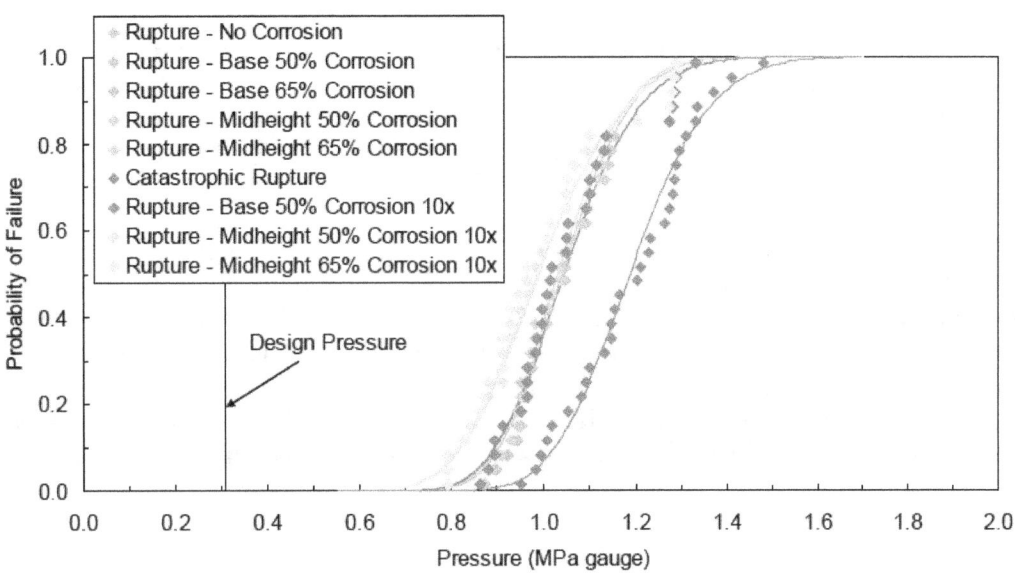

Figure 5.9: Fragility Curves for Rupture and Catastrophic Rupture

Figure 5.10 shows the cumulative probability of failure (leak) curves obtained in the Spencer study for the containment with corrosion degradation at the midheight and at the basemat, along with the fragility of the containment in its original condition. As would be expected from the baseline analysis, the drop in capacity due to corrosion at the mid-height is much greater than that due to corrosion at the basemat. The introduction of multiple regions of corrosion does not affect the cumulative probability of failure since it is assumed that when one corroded locations tears, all 10 regions tear simultaneously.

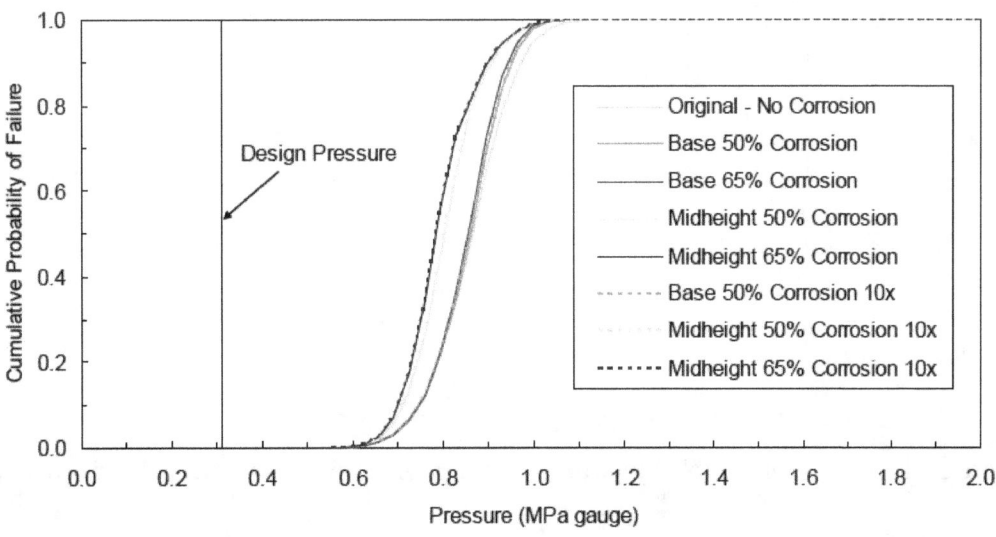

Figure 5.10: Cumulative Probability of Failure Curves for Original and Degraded RC Containment

The conditional probability of failure curves were computed for each of the eight cases and shown in Figure 5.11 through Figure 5.18. Since the rupture and catastrophic curves are relatively constant throughout the original five cases examined by Spencer et al. (2006), the conditional probability distributions are extremely similar. At low pressures ($P < 0.65$ MPa), the probability of leak is equal to 1 with the rupture and catastrophic rupture probabilities at 0. Increases in pressure cause the rupture and eventually the catastrophic rupture conditional probabilities to increase. The leak probability decreases quickly as the rupture probabilities increase. The rupture curve peaks at approximately 1.1 MPa. At this point, both the leak and rupture curves decrease gradually to probabilities of 0 and the catastrophic curve increase to a probability of 1 at approximately 1.7 MPa. For the 10 regions of basemat corrosion cases, the change in the conditional probabilities is similar to those for the original cases. However, the cases of multiple corrosion at the midheight do cause a noticeable increase in the rupture probabilities at lower pressures. The slight "wave" in the leak and rupture curves in Figure 5.17 and Figure 5.18 reflect the sensitivity to the curve fits used at the low pressure ($P < 0.75$ MPa) for the leak and rupture fragility curves in Figure 5.8 and Figure 5.9. This is especially true for those two cases where the 10 regions of midheight corrosion dominate the rupture areas at low pressures.

The fragility and conditional failure curves presented in this section are used as input in the PRA analyses. Those results are shown in Section 5.3. As mentioned in the previous section, the deterministic MELCOR/MACCS analyses were performed using three sets of data. The best estimate (or nominal case) uses the data from the previous section. The lower and upper bound sets of analyses use the structural analysis results from this section. The next section provides the area vs. pressure curves for each set of analyses as they were extracted from the structural analyses.

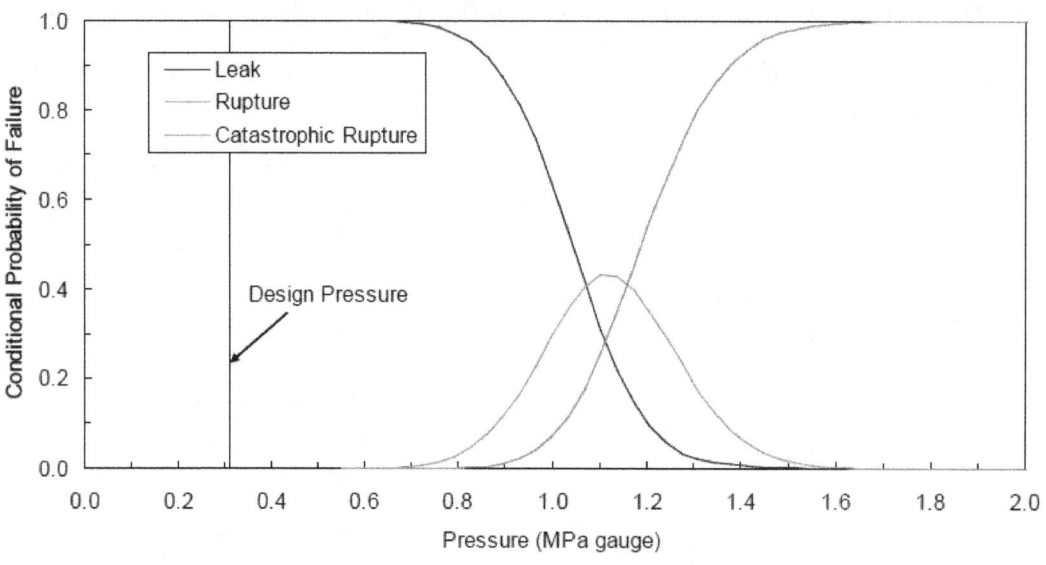

Figure 5.11: Conditional Probability of Failure for Original Containment (No Corrosion)

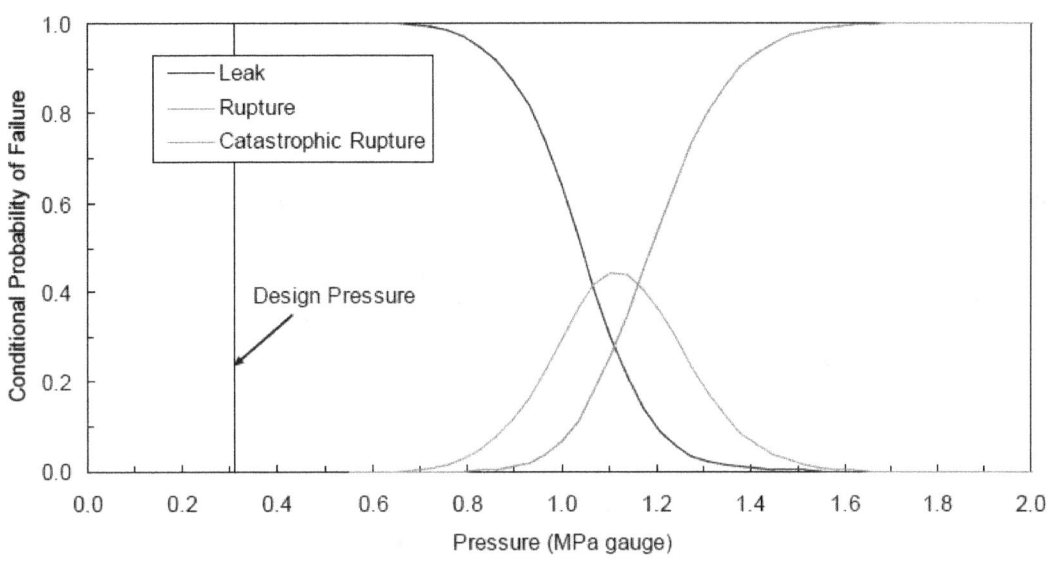

Figure 5.12: Conditional Probability of Failure for 50% Corrosion at Basemat

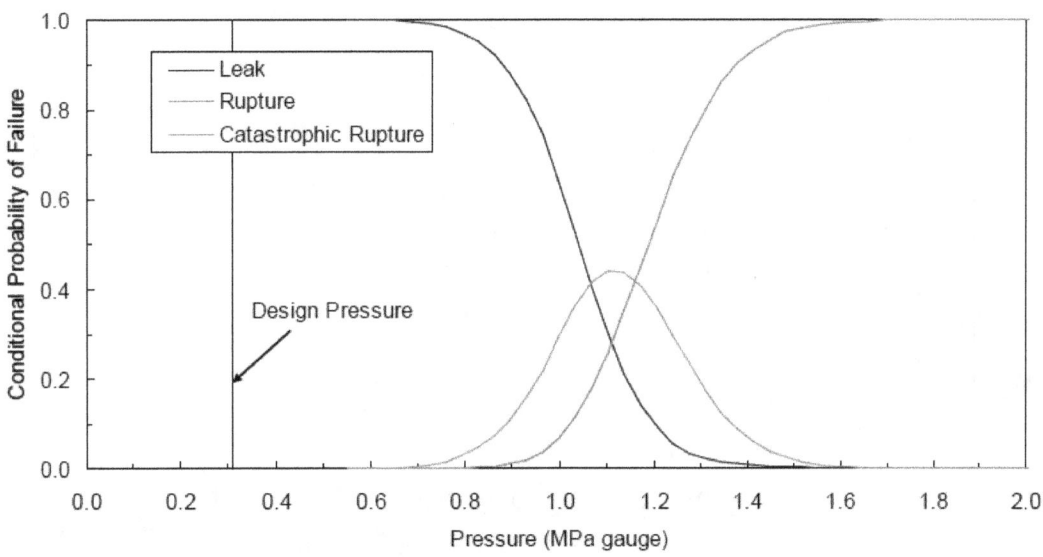

Figure 5.13: Conditional Probability of Failure for 65% Corrosion at Basemat

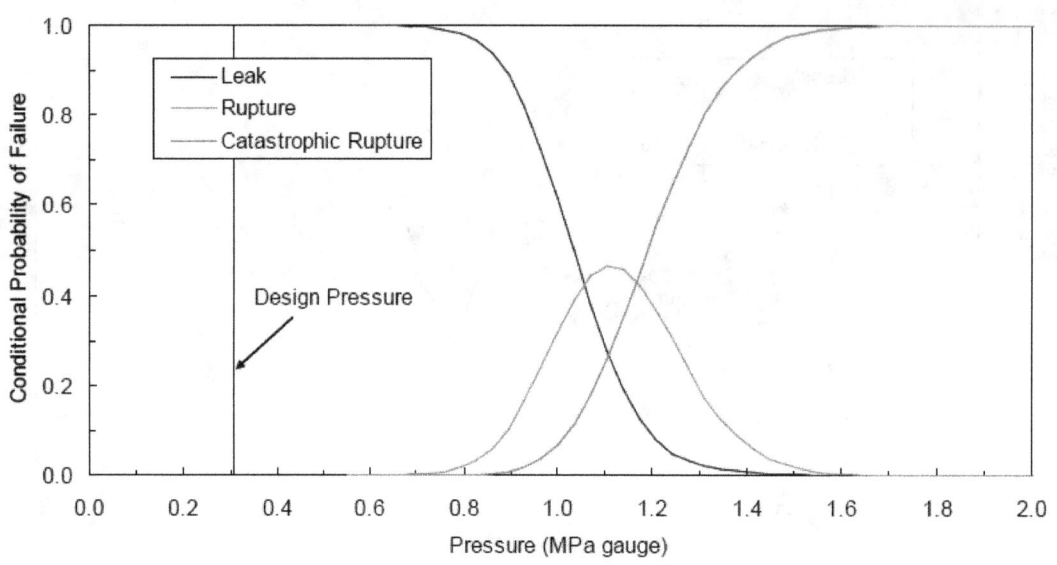

Figure 5.14: Conditional Probability of Failure for 50% Corrosion at Midheight

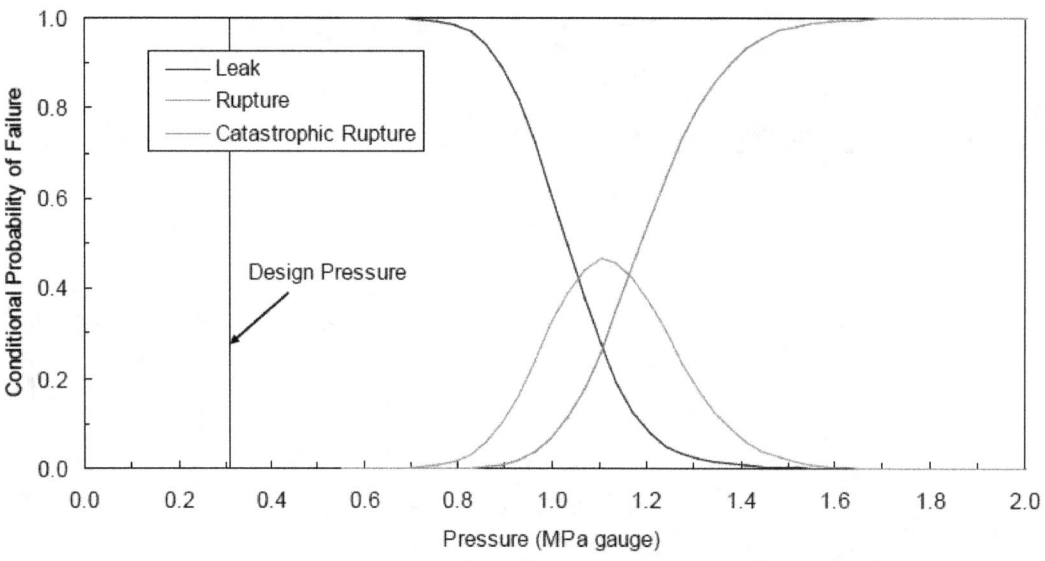

Figure 5.15: Conditional Probability of Failure for 65% Corrosion at Midheight

34

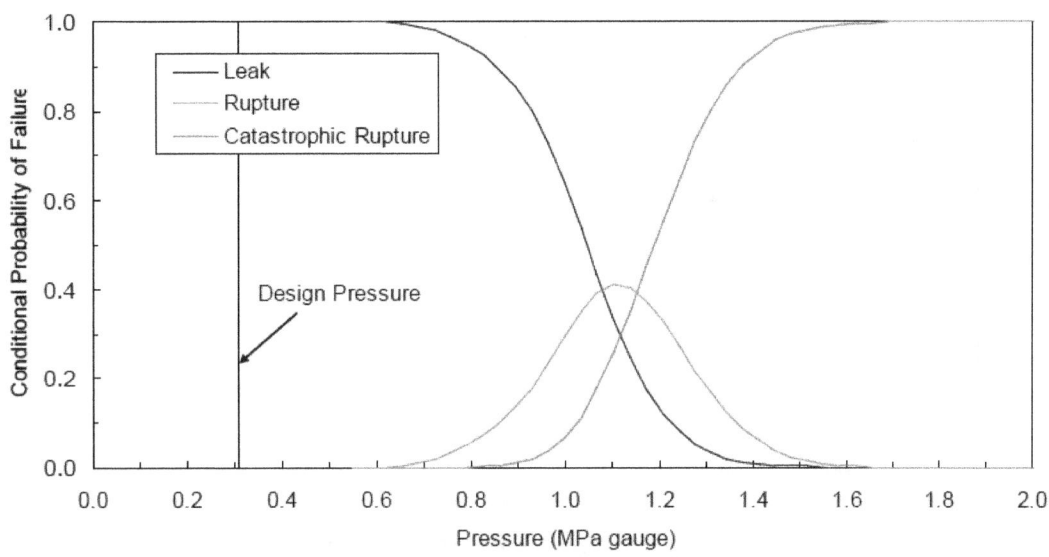

Figure 5.16: Conditional Probability of Failure for 10 Regions of 50% Corrosion at Basemat

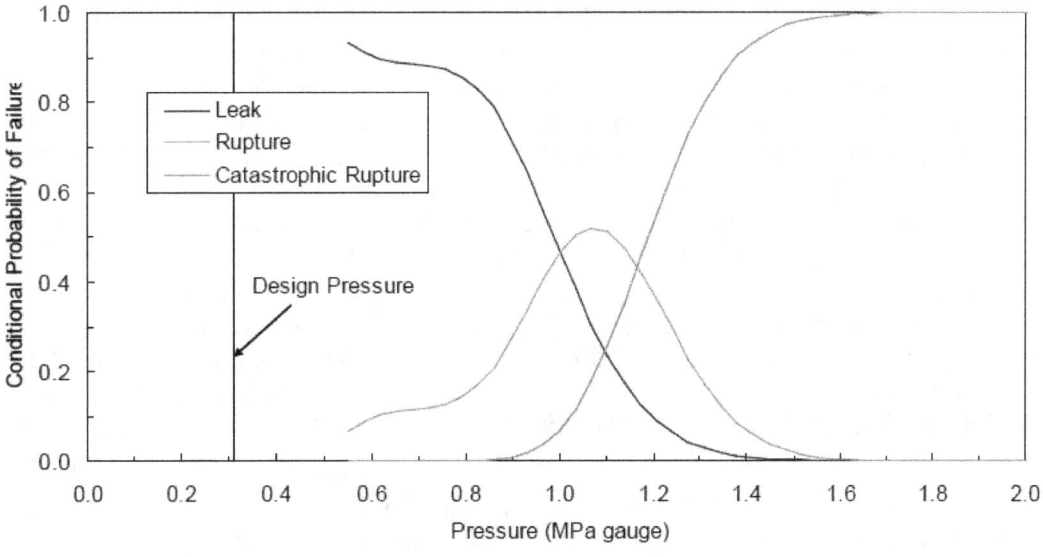

Figure 5.17: Conditional Probability of Failure for 10 Regions of 50% Corrosion at Midheight

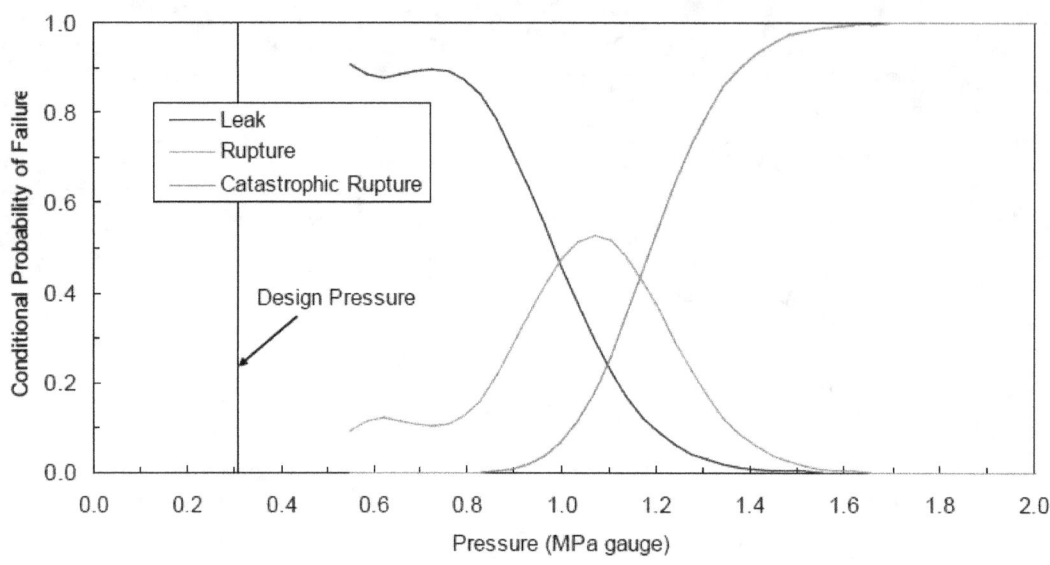

Figure 5.18: Conditional Probability of Failure for 10 Regions of 65% Corrosion at Midheight

5.1.5   Leak Area Curve Development for Deterministic Accident Analyses

As mentioned in the previous two sections, the MELCOR analyses require the area vs. pressure curves for the containment behavior as input. These curves were developed from the structural analyses for five different cases: no corrosion, 50% corrosion at the midheight, 65% corrosion at the midheight, 10 locations of 50% corrosion at the midheight, and 10 locations of 65% corrosion at the midheight. The basemat corrosion cases were initially examined as well, but they were not included since their area vs. pressure curves did not vary from the no corrosion case until well past the point where the MELCOR analyses showed peak pressures. This is not a general conclusion regarding basemat corrosion vs. other locations. The exact location, extent, level, and plant specific geometry all affect the severity of a given corrosion site and should be examined individually. In addition, curves were developed for each case for the best estimate structural analyses described in Section 5.1.3, and lower and upper bound structural analyses originally performed for the PRA analyses which are outlined in Section 5.1.4. The procedure in Section 2.1.1.3 was used to compute the ratio of the total rupture area, A_{total}, as a function of the pressure. This function was originally determined not just for the best estimate case, but also for each of the 30 analyses used in the fragility curve computations in the previous section. The best estimate total area versus pressure curves for each of the cases examined are shown in Figure 5.19 and Figure 5.20. Figure 5.20 illustrates the detailed view of the curves in the region of area and pressure that are obtained in the MELCOR analyses. For the lower and upper bound analyses, the area vs. pressure curves were developed by bounding the area vs. pressure curves for the 30 analyses used in the fragility analysis. The 30 curves for a given corrosion case were plotted together and lowest and highest (upper) most curves were drawn as to bound that space. Therefore, "lower bound" refers to the lower bound of the pressure for the 30 area vs. pressure curves. The "upper bound" therefore refers to the upper bound of the pressure for the 30 area vs. pressure curves. The lower bound curves are shown in Figure 5.21 and Figure 5.22. The upper bound curves are shown in Figure 5.23 and Figure 5.24. The cases and labeling used for the MELCOR and MACCS analyses are summarized in Table 5.3. It is important to identify the lower and upper bound identifiers with the lower and upper leak pressures and not lower and upper bound for consequences. This is critical since the lower bound cases lead to earlier leaks at lower pressures which can lead to higher consequences. The opposite is therefore true for the upper bound cases leaking at higher pressures and possibly leading to lower consequences. The next section on the MELCOR and MACCS analyses describe the consequence results in detail.

Table 5.3:   Corrosion Condition

Condition	Label
no corrosion	
lower bound	LB1
nominal	N1
upper bound	UB1
1 location, 50% midheight corrosion	
lower bound	LB2
nominal	N2
upper bound	UB2
1 location, 65% midheight corrosion	
lower bound	LB3
nominal	N3
upper bound	UB3
10 locations, 50% midheight corrosion	
lower bound	LB4
nominal	N4
upper bound	UB4
10 locations, 65% midheight corrosion	
lower bound	LB5
nominal	N5
upper bound	UB5

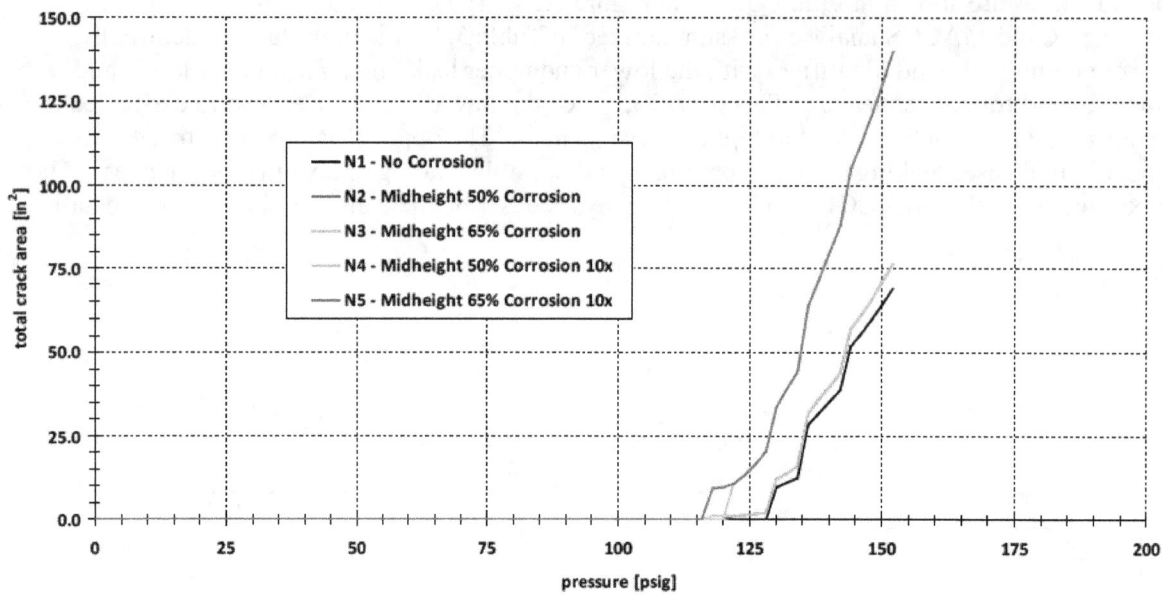

Figure 5.19: Total Crack Area vs. Pressure for Best Estimate Cases (Nominal)

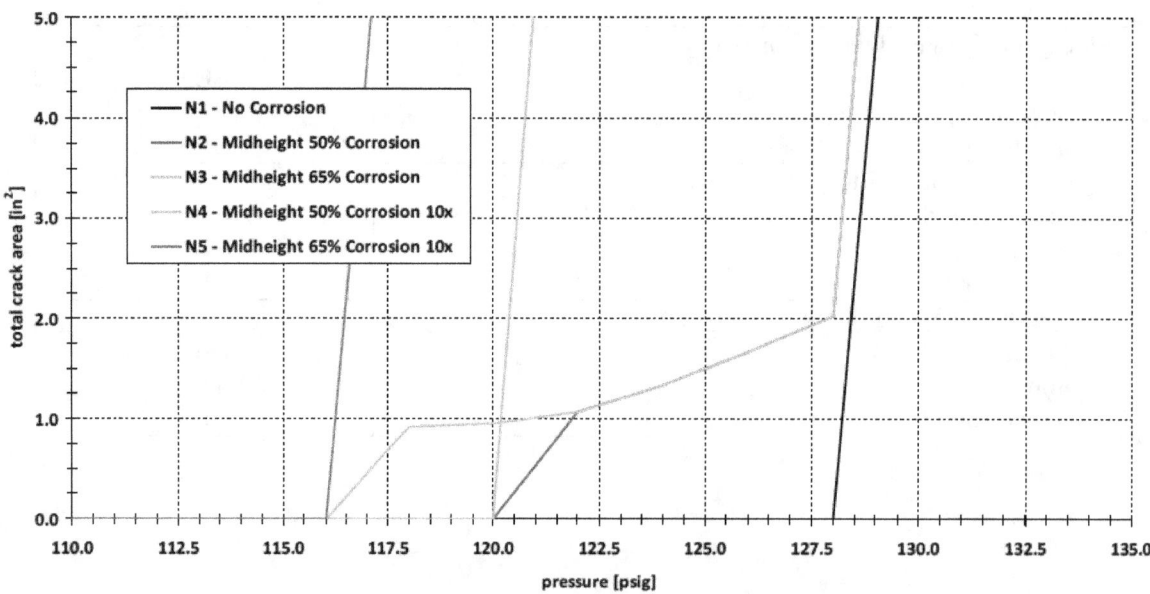

Figure 5.20: Total Crack Area vs. Pressure Detailed View for Best Estimate Cases (Nominal)

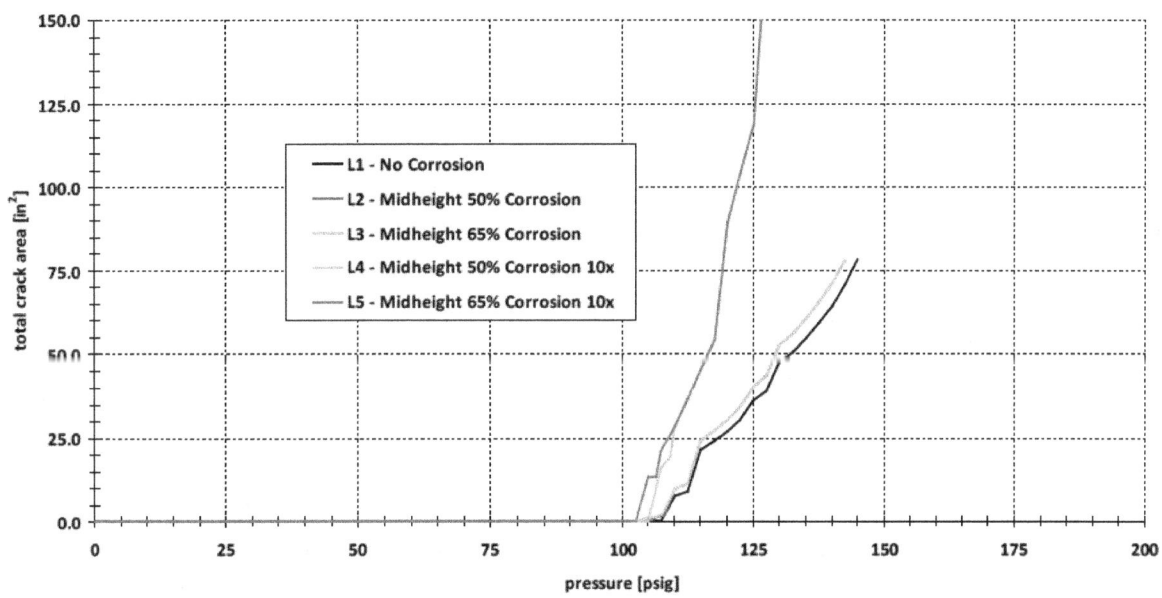

Figure 5.21: Total Crack Area vs. Pressure for Lower Bound Cases

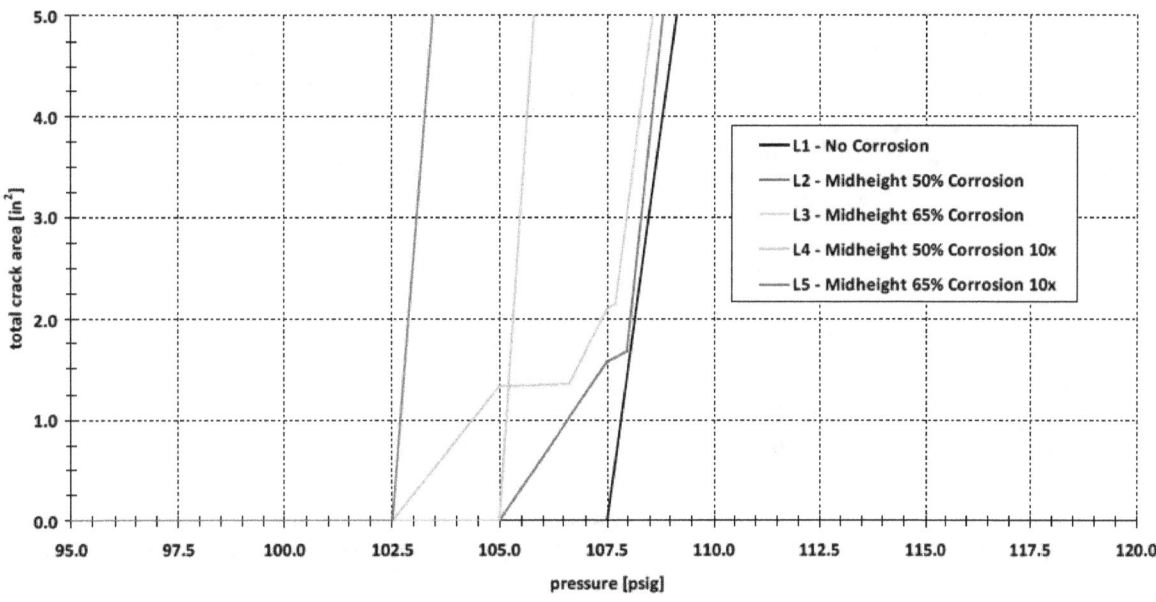

Figure 5.22: Total Crack Area vs. Pressure Detailed View for Lower Bound Cases

39

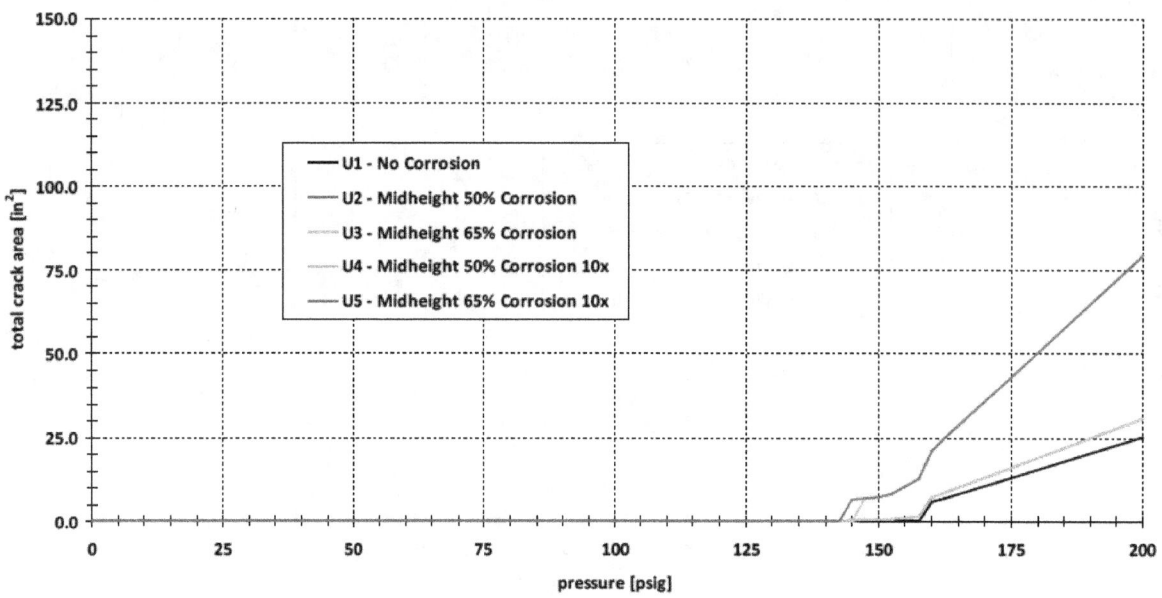

Figure 5.23: Total Crack Area vs. Pressure for Upper Bound Cases

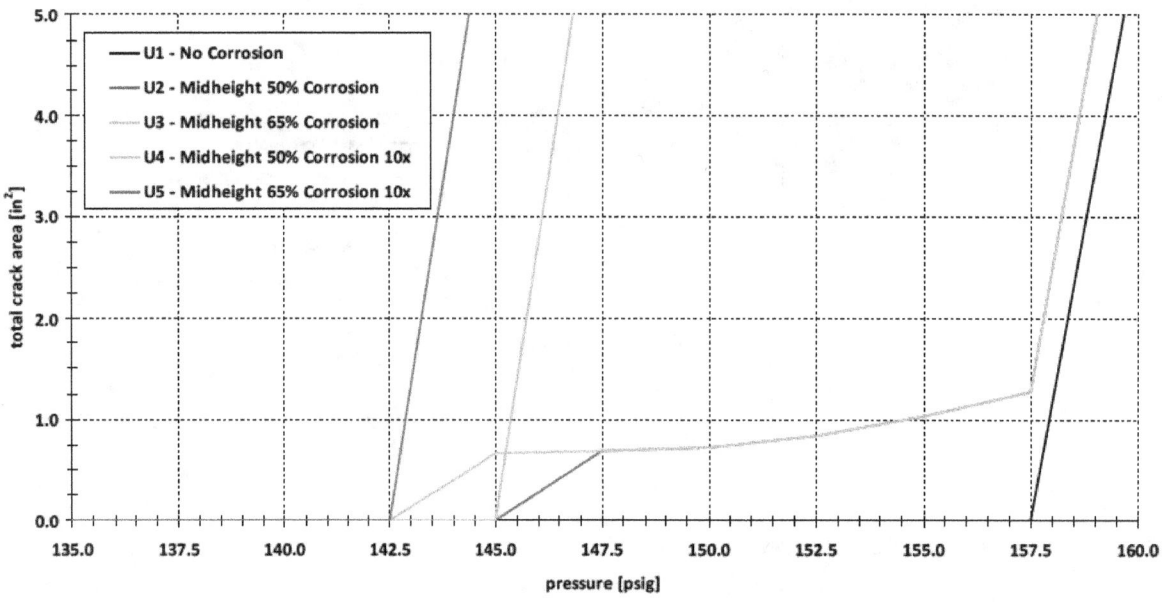

Figure 5.24: Total Crack Area vs. Pressure Detailed View for Upper Bound Cases

5.2 MELCOR and MACCS Deterministic Accident Analysis

5.2.1 MELCOR Model for a 3-Loop PWR with a Reinforced Concrete Subatmospheric Containment

A model of the Surry Power Station units has been developed by Sandia National Laboratories using MELCOR (Gauntt, et al., 2005, U.S.NRC, 2010). The model was originally generated for code assessment applications at Idaho National Engineering Laboratories (INEL) in 1988 (Dobbe, 1988). This model has been updated by Sandia National Laboratories (1990 to present) for the purpose of testing new models, advancing the state-of-the-art in modeling of PWR accident progression, and providing support to decision-makers at the U.S. Nuclear Regulatory Commission (NRC) for analyses of various issues that may affect operational safety. Significant changes have been made during the last ten years in the approach to modeling core behavior and core melt progression, as well as the nodalization and treatment of coolant flow within the RCS and reactor vessel. Assessment reports have been prepared to discuss this model evolution as part of the MELCOR code development program (Gauntt, et al. 2005).

Recent significant enhancements to the MELCOR model have been made to take advantage of MELCOR 1.8.6 modeling capabilities. The updates concern two key areas. The first area is an upgrade to MELCOR Version 1.8.6 vessel and core modeling. These enhancements include:

- a hemispherical lower head model that replaces the flat bottom-cylindrical lower head model,
- new models for the core former and shroud structures that are fully integrated into the material degradation modeling, including separate modeling of debris in the bypass region between the core barrel and the core shroud,
- models for simulating the formation of molten pools both in the lower plenum and the upper core, crust formation, convection in molten pools, stratification of molten pools into metallic and oxide layers, and partitioning of radionuclides between stratified molten pools,
- a reflood quench model that separately tracks the component quench front, quench temperature, and unquenched temperatures,
- a control rod silver aerosol release model, and
- an application of the CORSOR-Booth release model for modern high-burn-up fuel.

The second area focused on the addition of user-specified models to represent a wide spectrum of plant design features and safety systems to broaden the capabilities of MELCOR to a wider range of severe accident sequences. These enhancements included:

- update of the containment leakage model,
- update of core degradation modeling practices,
- modeling of individual primary and secondary relief valves with failure logic for rated and degraded conditions,
- update of the containment flooding characteristics,
- heat loss from the reactor to the containment,
- separate motor and turbine-driven auxiliary feedwater models with control logic for plant automatic and operator cooldown responses,
- new turbine-driven auxiliary feedwater models for steam flow, flooding failure, and performance degradation at low pressure,
- nitrogen discharge model for accumulators,
- update of the fission product inventory, the axial and radial peaking factors, and an extensive fission product tracking control system, and

41

- improvements to the natural circulation in the hot leg and steam generator and the potential for creep rupture.

5.2.2 Description of MELCOR Model Modifications for Degraded Containment Analysis

In the MELCOR model, containment failure is simulated with a flow path[1] whose area is a function of the containment pressure. Figure 5.25 shows the nodalization of the containment with the location of the containment failure flow path.

A MELCOR flow path is defined by the following parameters:

Orientation: Based on the geometry of the problem the containment failure flow path was modeled as a horizontal, rather than a vertical, flow path.

Opening Height: The location of the containment failure is between control volume 5 (CV5) and the Environment (CV999). Since the source term release only consists of vapors and gases, as long as the containment failure flow path connects the Basemat CV to the Environment CV, the specific opening height location is not important (all of the leaks we consider are combined in CV5, so a detailed breakdown of the different leak paths will not affect the source term). As such, a height half-way between the top of the basemat and the bottom of the upper dome was selected (16.85 m, 55.28 ft).

Open Area: The open area is set equal to the maximum open area for the area vs. pressure relationship of interest. As the pressure increases, the open area from the area vs. pressure curves is used. At the point that the pressure decreases, the open area in the MELCOR analyses remains at the maximum value reached. This is a slightly conservative assumption, since there will be some crack closure. However, the crack region will now be irreversibly damaged and would not follow back along the original curve. In the absence of more sophisticated techniques, the area is assumed to remain at its maximum value as the pressure drops.

Length: The flow path length was assumed to be the thickness of the containment wall (4.5 ft). While the cracks may actually have a longer length, this minimum value was chosen as a conservative lower bound in the sense that a shorter length corresponds to a higher flow rate, and hence more source term release from the flow path.

Fraction of Flow Path Open: The fraction of flow path that is open is derived from the area vs. pressure relationship of interest by dividing the area by the maximum open area.

Forward and Reverse Loss Coefficients: The code default values of 1.0 were used.

Forward and Reverse Choked Flow Coefficients (i.e., *k*): The code default values of 1.0 were used.

---

[1] A flow path is a MELCOR model element that connects two control volumes (CVs) (in this particular case, the containment CV to the environment CV) and allows material flow (e.g., liquid, vapor, gas, source term aerosol) between the CVs.

Hydraulic Diameter: The hydraulic diameter was computed assuming a set of 12 identical cracks. Each crack has a rectangular geometry. The in-plane length of each crack is assumed to be 2 ft. Although the number or cracks and lengths varied in the computation of the area vs. pressure curves from the structural analyses, the numbers were held constant in the MELCOR analysis when computing the hydraulic diameter.

The hydraulic diameter is defined as

$$D_h = 4\frac{A}{P} \qquad (5.5)$$

where

D_h - hydraulic diameter
A - flow path area
P - flow path perimeter

MELCOR requires that a constant value be specified for the hydraulic diameter. Therefore, the actual temporal change in hydraulic diameter as the crack opens cannot be modeled. As a conservatism, the hydraulic diameter was calculated based on the maximum open area. Hence given the assumption of a rectangular crack geometry and "N" cracks, the perimeter of the containment failure flow path is equal to

$$P = 2N(h + w) \qquad (5.6)$$

$$w = \frac{A}{Nh} \qquad (5.7)$$

where

N - number of cracks
h - crack height
w - crack width

This treatment yields the maximum possible hydraulic diameter, which maximizes the flow rate, and hence source term release from the flow path. In addition, it should be noted that the hydraulic diameter is not a function of the crack length (e.g. containment wall thickness). It is only a function of the crack open area and the perimeter of the crack opening.

Surface Roughness: A surface roughness (0.3 mm) similar to a lower-bound value for concrete pipes was assumed. This value was chosen as a conservative lower bound in the sense that a lower surface roughness corresponds to a higher flow rate, and hence a higher source term release from the flow path.

It should be noted that for the majority of the time that leakage occurs, the flow through the leak path will be choked. In this flow regime k, D_h, W, P, and surface roughness are not relevant. However, regardless of the flow regime (i.e., choked or non-choked) MELCOR requires that these parameters be specified. Moreover, the parameters are relevant during the time that the flow through the leak path is non-choked.

For this analysis, the containment failure flow path in the original model was replaced with a set of 15 flow paths (see Table 5.3), each of which simulates one of the postulated containment failure area versus containment pressure response curves (see Figure 5.19 through Figure 5.24). Control logic was added to the model which allows the containment failure flow paths to be either "active" or "inactive". The fraction open area for an active flow path is set equal to that derived from its failure area versus containment pressure response curve. The fraction open area for an inactive flow path is equal to zero – hence no flow occurs through an inactive flow path. When the model was run, only one of 15 containment failure flow paths would be set to "active", with the remaining 14 set to "inactive".

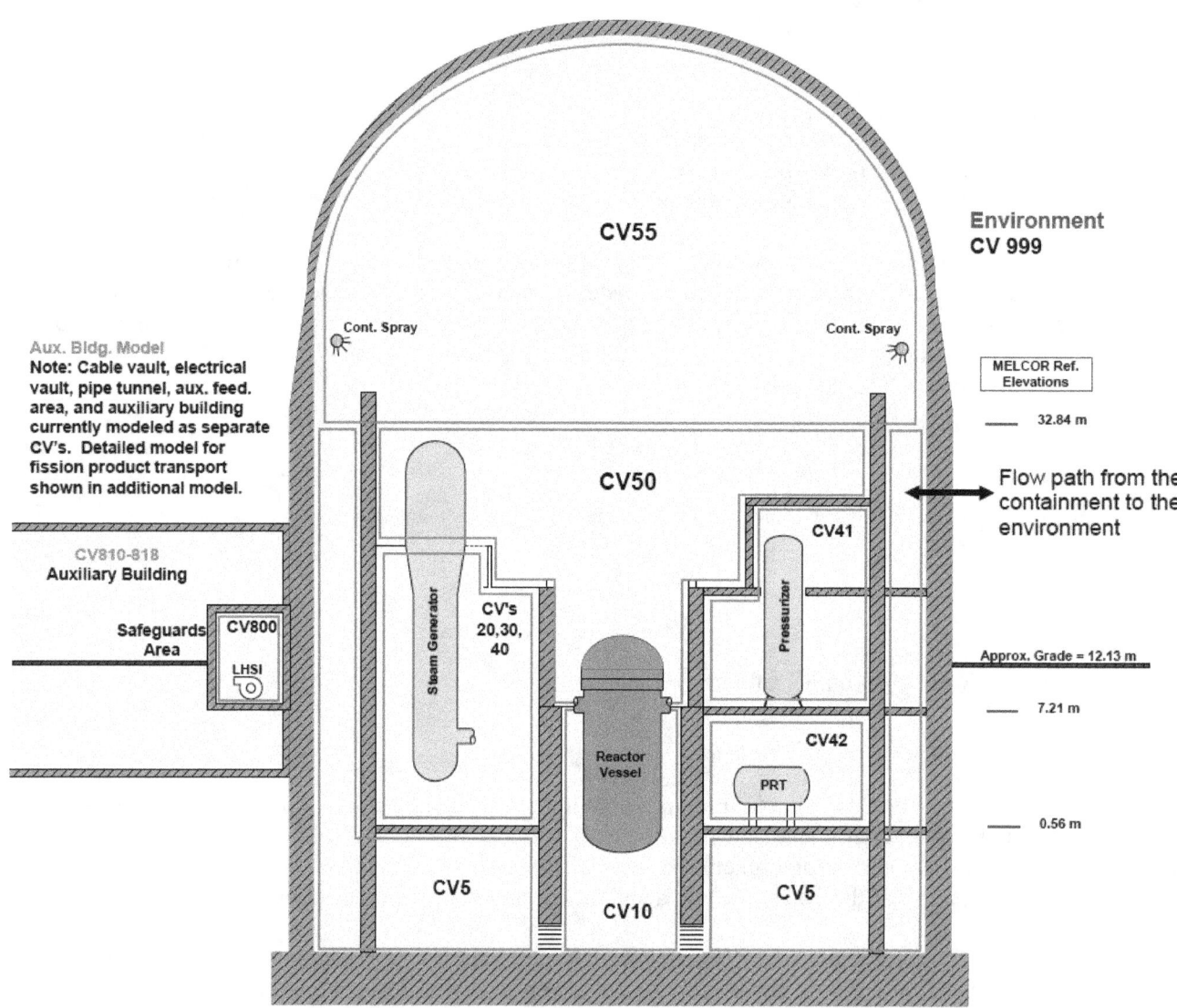

Figure 5.25: Containment Nodalization

5.2.3 Description of Degraded Containment MELCOR Calculations

For the purposes of quantifying variances in consequences of liner degradation, two unmitigated station blackout (SBO) sequences were selected; the unmitigated short-term station blackout (STSBO) and the unmitigated long-term station blackout (LTSBO). The SBO sequences do not credit any successful mitigative actions, such as those developed in the severe accident management guidelines (SAMGs) and mitigation measures codified in 10 CFR 50.54(hh). The unmitigated SBO severe accident sequences produce an over-pressurization failure challenge to the containment, which allows quantification of the source term as a function of liner degradation parameters. The range of liner degradations are not best-estimate representations of the reference plant (i.e., Surry) but rather parameters to evaluate regulatory issues related to liner degradation. The resulting distributions of results are evaluated to identify trends as a function of degradation type and magnitude as well as identify threshold conditions that lead to a significantly different result.

The short-term station blackout is initiated by a large earthquake (0.5–1.0 peak ground acceleration - pga). The seismic event results in a loss-of onsite power (LOOP) and failure of onsite emergency alternating current (AC) power resulting in a station blackout (SBO) event where neither onsite nor offsite AC power are recoverable. All systems dependent on AC power are unavailable, including the containment engineering safety systems (e.g., the containment spray and fan coolers). The seismic event also causes a loss of direct current (DC) power, which makes the turbine-driven auxiliary feedwater (TDAFW) system unavailable. The reactor coolant system (RCS) and containment are undamaged and the containment is isolated. No instrumentation is available. Following the loss of the seal cooling flow, the reactor coolant pump (RCP) seals will nominally leak at 21 gpm (i.e., at normal operating pressure and temperature). The RCP seals fail later in the accident when the RCP seal region heats to saturated conditions. All emergency core cooling system (ECCS) equipment is inoperable primarily due to electrical and physical system damage due to the seismic event. However, the accumulators are available when the RCS depressurizes.

The LTSBO is initiated by a moderately large earthquake (0.3–0.5 pga). The seismic event results in LOOP and failure of onsite emergency AC power resulting in a station blackout event where neither onsite nor offsite AC power are recoverable. Unlike the unmitigated STSBO described above, the turbine-driven auxiliary feedwater (TDAFW) system is available initially. In the long term, the loss of the TDAFW may occur due to battery depletion and loss of DC power for sensing and control. The TDAFW is also only available until the emergency condensate storage tank empties. The station batteries provide instrumentation until they exhaust at 8 hr. The secondary power-operated relief valves (PORVs) are initially available for a manual 100°F/hr system cooldown. The secondary PORVs are assumed to close following battery failure. No other systems are available.

Although the LTSBO is not successfully mitigated by emergency procedures and equipment codified in 10 CFR 50.54(hh), it is assumed that two operator actions are successful. The operator is (a) initially successful to maintain normal water levels in the steam generators (SG) using TDAFW and (b) performs a controlled depressurization of the steam generators to approximately 120 psi to achieve an RCS cooldown of 100°F per hour by manually opening the SG PORVs.

At the start of the accident progression in the unmitigated STSBO sequence, the reactor successfully scrams in response to the loss of power. The main steam line isolation and

45

containment isolation valves close in response due to the loss of power. The reactor coolant and main feedwater pumps also trip to the loss of power. Once the main steam lines close, the normal mechanism of heat removal from the primary system is unavailable. Consequently, both the primary and secondary system pressures rise. The secondary system quickly pressurizes to the safety relief valve opening pressure, which results in the safety relief valves to open and then subsequently close when the closing pressure criterion is achieved. The relief flow through the SG safety relief valves is the principal primary system energy removal mechanism in the first hour. The water inventory in the steam generators was completely boiled away by 1 hr 16 min. Although the steam generator relief valves continue to cycle and release steam, the associated heat removal is inadequate and the primary system sharply increases to the pressurizer safety relief valve opening pressure. The safety valves on the pressurizer begin opening and closing to remove excess energy. However, the pressurizer relief valve flow causes a steady decrease in the primary system coolant inventory. The fuel starts to uncover at 2 hr 19 min. (See Table 5.4 for a summary of timings in the unmitigated STSBO).

Following the uncovery of the fuel, an in-vessel natural circulation flow develops between the hot fuel in the core and the cooler structures in the upper plenum. Hot gases from inside the vessel also flow along the top the hot leg and into the steam generator. The large masses of the hot leg nozzle, hot leg piping, and the steam generator tubes absorb the heat from the gases exiting the vessel. The cooler gases leaving the steam generator return to the vessel along the bottom of the hot leg. Due to its close proximity to the hot gases exiting in the vessel, the hot leg nozzle at the carbon steel interface region to the stainless steel piping was predicted to fail by creep rupture at 3 hr 45 min.

Upon creep failure of the hot leg nozzle, a large hole opened that rapidly depressurized the RCS (i.e., like a large break loss-of-coolant accident). The RCS depressurization permitted a complete accumulator injection at low-pressure. Following the accumulator injection, the decay heat from the fuel boiled away the injected water. By 4.3 hr, a large debris bed had formed in the center of the core. The debris continued to expand until 5.8 hr when all the fuel had collapsed and was resting on the core plate. The hot debris failed the core support plate and fell onto the lower core support plate, which failed at 6.6 hr. Following the lower core support plate failure, the debris bed relocated onto the lower head. The small amount of remaining water in the lower head was quickly boiled away. The hot debris quickly heated the lower head and it failed at 7 hr 16 min due to the creep rupture failure criterion (i.e., primarily due the thermal stress component due to the low differential pressure).

Nearly all the hot debris relocated from the vessel by 7.5 hr into the reactor cavity in the containment under the reactor vessel. The hot debris boiled away the water in the reactor cavity and started to ablate the concrete. The ex-vessel core-concrete interactions (CCI) continued for the remainder of the sequence, which generated non-condensable gases. In addition, the hot gases exiting the reactor cavity and the radioactive heating from airborne and settled fission products steadily evaporated the water on the containment floor outside the reactor cavity from 7.3 hr to 44 hr. The resultant non-condensable gas and steam generation pressurized the containment. At 25.5 hr, the containment failed due to liner tearing near the containment equipment hatch at mid-height in the cylindrical region of the containment (in the original MELCOR model, the location and timing varies slightly for each of the degradation cases). The containment continues to pressurize until the leakage flow balanced the steam and non-condensable gas generation. By 44 hr, all the water on the floor has evaporated. The containment depressurized thereafter due to only a smaller gas loading from the non-condensable gas generation. The containment failure locations were all placed at locations that would not lead to releases into surrounding buildings (e.g., not adjacent to the auxiliary or safeguards buildings). Consequently, all released fission products are released directly to the environment.

The fission product releases from the fuel started following the first thermo-mechanical failures of the fuel cladding in the hottest rods at 2 hr 57 min, or about 38 min after the uncovery of the top of the fuel rods. The in-vessel fission product release phase continued through vessel failure at 7.3 hr. Approximately 97% and 98% of the iodine and cesium, respectively, were released from the fuel prior to vessel failure while the remaining amount was released ex-vessel. At the time of the hot leg failure, approximately 40% of these volatile radionuclides had been released. The resultant blowdown of the vessel immediately discharged the airborne fission products to the containment. Within the first day, most of the airborne fission products in the containment settled on surfaces. This was significant because the containment failure occurred at 25 hr 32 min. Consequently, there was little airborne mass that could be released to the environment. Again, these times are those taken from the original MELCOR accident scenario. This study uses different containment capacity models, and therefore, the time for increased containment leakage and subsequent events in the accident progression will vary slightly for each of the degradation cases.

Table 5.4:   STSBO Accident Progression Sequence of Events

Event Description	Time
Initiating event	00:00:00
MSIVs close	00:00:00
First SG SRV opening	00:03:00
SG dryout	01:16:00
Pressurizer SRV opens	01:27:00
Pressurizer relief tank rupture disk opens	01:46:00
Start of fuel heatup	02:19:00
RCP seal failures	02:45:00
First fission product gap releases	02:57:00
Creep rupture failure of the A loop hot leg nozzle	03:45:00
Accumulators start discharging	03:45:00
Accumulators are empty	03:45:00
Vessel lower head failure by creep rupture	07:16:00
Debris discharge to reactor cavity	07:16:00
Cavity dryout	07:27:00
Containment at design pressure (0.31 MPa, 45 psig)	11:00:00
Start of increased leakage of containment (P/P_{design} = 2.18)	25:32:00
Containment pressure increase slows	32:00:00
Containment pressure stops decreasing	44:14:00
End of calculation	4 days

The response of the unmitigated LTSBO is similar to the unmitigated STSBO. (See Table 5.5 for a summary of timings in the unmitigated LTSBO). However, the key events in the LTSBO are delayed due the operation of the LTSBO and the two successful operator actions. The TDAFW initiates at full flow but is subsequently controlled by the operator after 15 min to maintain level. The TDAFW restores the steam generator liquid levels by about 30 min and is throttled thereafter. After the closure of the main steam isolation valves, the secondary system quickly pressurizes to the safety relief valve opening pressure, which causes the safety relief valves to open and then subsequently close when the closing pressure criterion is achieve. The relief flow through the SG SRVs is the principle primary system energy removal mechanism in the first 90 min. At 90 min, the operator starts a controlled (~100°F/hr) cooldown of the primary system by venting the steam generator PORVs. As the secondary pressure decreases, the saturation temperature of the water in boiler section of the steam generator also decreases, which cools the primary system fluid. At about 3.5 hr, the steam generators reached 0.93 MPa (120 psig), where the secondary system pressure was stabilized. No operator actions were credited to replenish the emergency condensate storage tank inventory (e.g., the source of water for the TDAFW) and it drains by 5 hr 8 min. Subsequently, the steam generator level starts to decrease and is empty by 12 hr 18 min.

At 8 hr, the station batteries were estimated to fail. At the same time, the steam generator relief valves closed and were no longer actively controlled. In response to the steam generator valve closure, both the primary and secondary systems rapidly pressurized. Once the steam generators boiled dry, the primary system pressurized to the pressurizer safety relief valve opening set point and began to relieve steam and water. The top of the fuel was uncovered by 14.3 hr and the core heatups began.

Similar to the STSBO (see above), natural circulation flows developed between the hot fuel in the core and the cooler structures in the upper plenum, hot leg, and the steam generator. The hot leg nozzle at the carbon steel interface region to the stainless steel piping was first predicted to fail by creep rupture at 17 hr 6 min. The remainder of the sequence of events were similar to the STSBO but delayed in timing as summarized in Table 5.5. The containment failed at 45.5 hr, due to liner tearing near the containment equipment hatch at mid-height in the cylindrical region of the containment (for the original MELCOR containment model).

The fission product releases from the fuel started following the first thermo-mechanical failures of the fuel cladding in the hottest rods at 16 hr 4 min, or about 1 hr 40 min after the uncovery of the top of the fuel rods. Approximately 99% of the iodine and cesium were released from the fuel prior to vessel failure while the remaining amount was released ex-vessel. The majority of the radionuclides released in-vessel went to the containment. Within 36 hr, most of the airborne fission products in the containment settled on surfaces. This was significant because the containment failure occurred at 45 hr 32 min. Consequently, there was little airborne mass that would be released to the environment. As with the STSBO, these times are for the original MELCOR models for containment capacity. The models developed in this study cause a slightly different timing for the events in the accident progression.

Table 5.5: LTSBO Accident Progression Sequence of Events

Event Description	Time
Initiating event	00:00:00
MSIVs close	00:00:00
TD-AFW auto initiates at full flow	00:00:30
Operators control TD-AFW to maintain level	00:15:00
Operators initiate controlled cooldown of secondary at ~100°F/hr	01:30:00
Vessel water level drains into upper plenum	01:57:00
Initial minimum vessel water level	02:30:00
Accumulators begin injecting	02:25:00
SG cooldown stopped at 120 psig to maintain TDAFW flow	03:35:00
First SG SRV opening	00:03:00
Emergency CST empty	05:08:00
dc Batteries Exhausted	08:00:00
S/G PORVs reclose	08:00:00
Pressurizer SRV opens	13:06:00
PRT failure	13:40:00
Start of fuel heatup	14:16:00
RCP seal failures (calculated)	14:46:00
First fission product gap releases	16:04:00
Creep rupture failure of the C loop hot leg nozzle	17:06:00
Accumulator empty	17:06:00
Vessel lower head failure by creep rupture	21:08:00
Debris discharge to reactor cavity	21:08:00
Cavity dryout	21:16:00
Containment at design pressure (0.31 MPa, 45 psig)	28:00:00
Start of increased leakage of containment (P/P$_{design}$ = 2.18)	45:32:00
Calculation terminated	4 days

The 15 corrosion conditions were evaluated for both the STSBO and LTSBO scenarios (see Table 5.3). Figure 5.26 through Figure 5.31 show the containment pressure as a function of time for each corrosion case and for the best estimate and lower/upper bounds. The release constituents are also computed in MELCOR for each of these cases to be used as input in the MACCS analyses described in the next section.

For a number of cases in Figure 5.26 through Figure 5.31, the pressure curves rise above or drop below other curves. This is due to the complex interaction between the pressure vs. area curve and the timing of the accident scenario being examined.

49

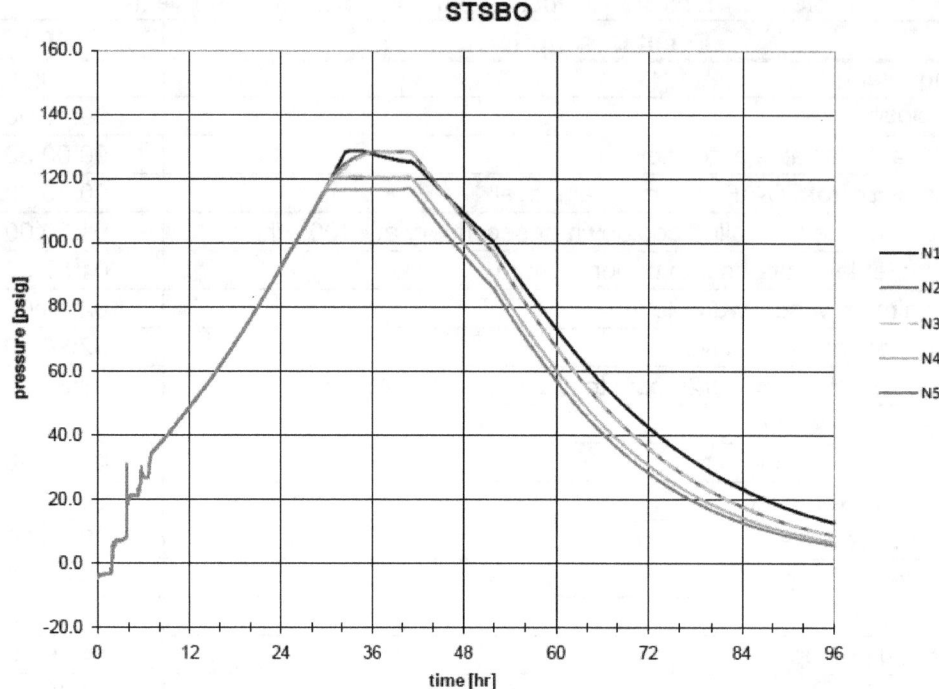

Figure 5.26: Containment Pressure – STSBO, Nominal Cases (see Table 5.3 for case descriptions)

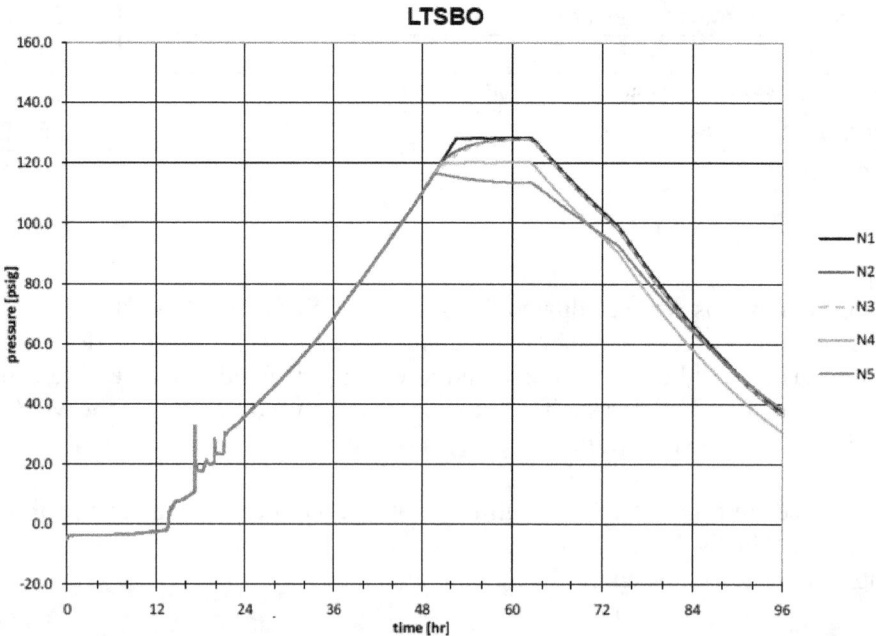

Figure 5.27: Containment Pressure – LTSBO, Nominal Cases (see Table 5.3 for case descriptions)

50

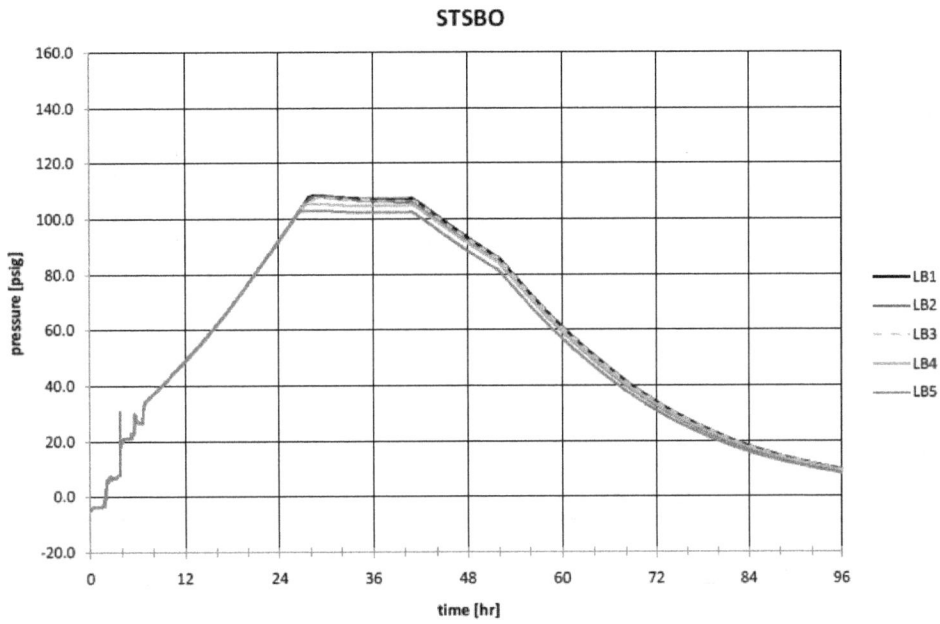

Figure 5.28: Containment Pressure – STSBO, Lower Bound Cases (see Table 5.3 for case descriptions)

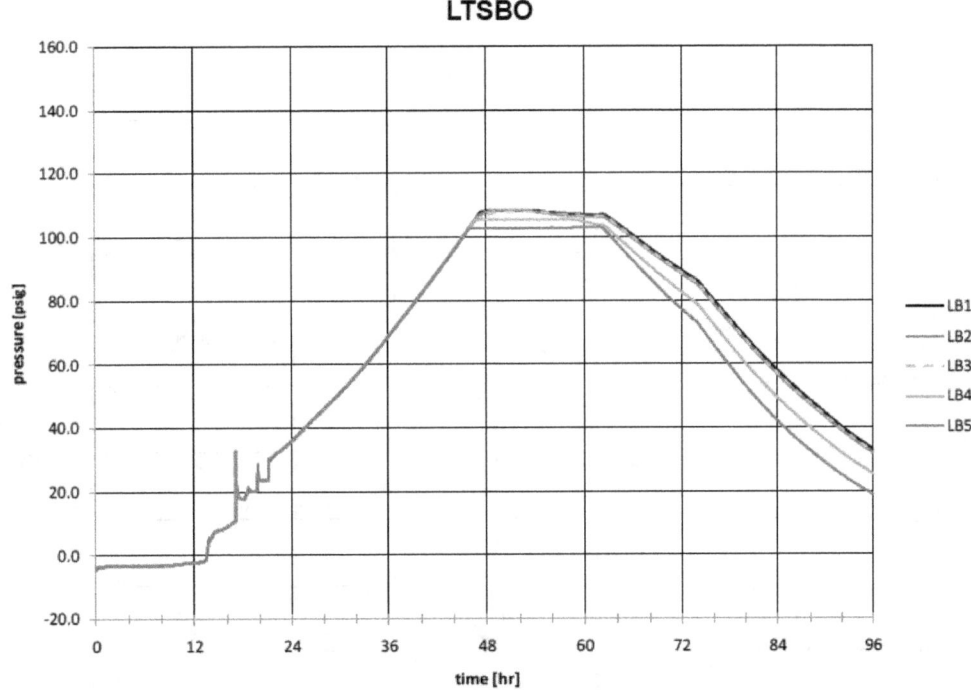

Figure 5.29: Containment Pressure – LTSBO, Lower Bound Cases (see Table 5.3 for case descriptions)

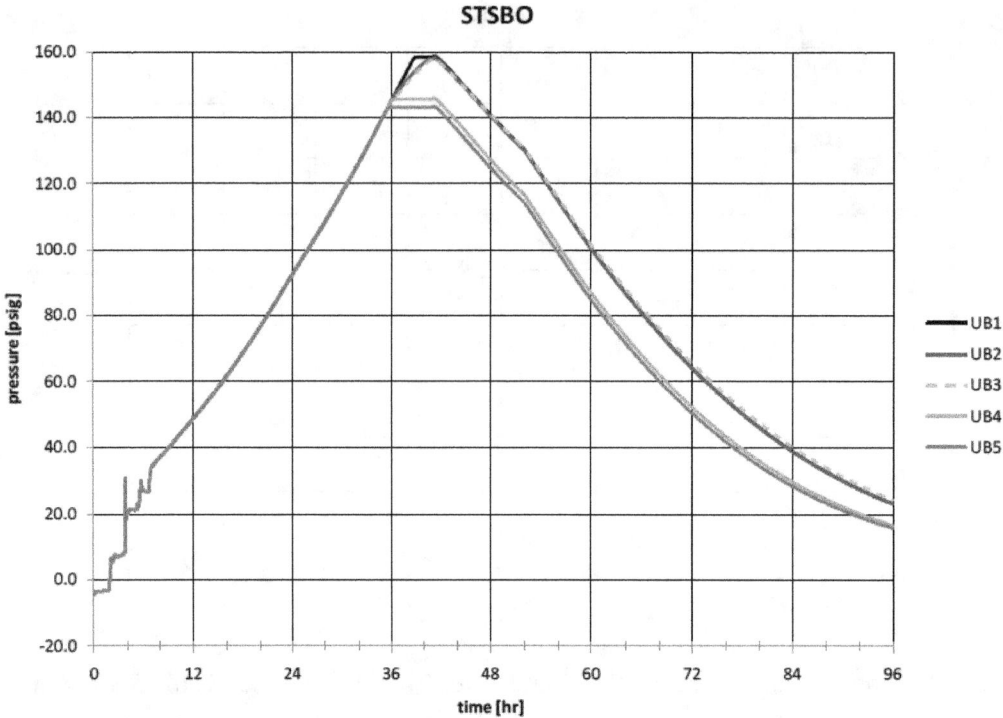

Figure 5.30: Containment Pressure – STSBO, Upper Bound Cases (see Table 5.3 for case descriptions)

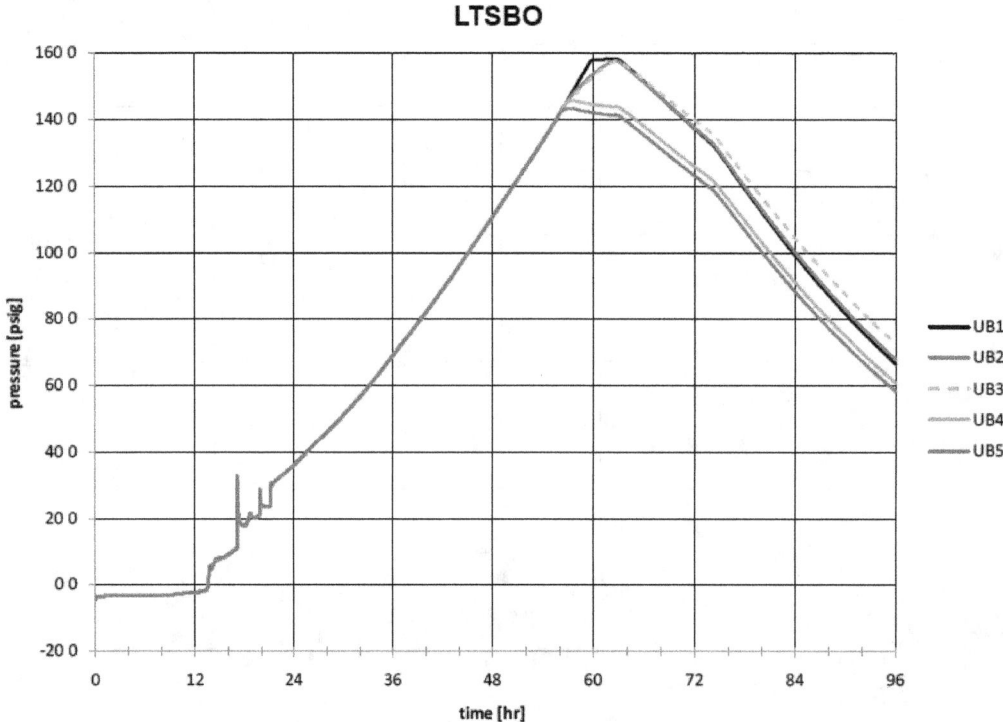

Figure 5.31: Containment Pressure – LTSBO, Upper Bound Cases (see Table 5.3 for case descriptions)

5.2.4 MACCS2 Analyses

The offsite consequence analysis was performed using MACCS2 for each of the source terms generated by MELCOR. The MACCS2 input model employs site specific emergency response scenarios for severe accidents that are characterized with the timing of actions taken by onsite and offsite response organizations to protect the general public by evacuation and sheltering. These emergency response scenarios were developed in order to achieve a greater degree of realism than previous consequence analyses performed by MACCS2. The following subsections summarize the development of those emergency response scenarios as well as generic and site specific input parameters used in this study.

5.2.5 MACCS2 Generic Input Parameters

Generic input parameters are neither site nor accident scenario specific. These include dose factors which are used to relate the dose to each organ to the concentration of each of the isotopes, specification of spatial grid elements to specify radial distances from the plant at which the results are to be reported, decontamination parameters, and some parameters concerning deposition and dispersion of released materials.

Many of these generic input parameters used in this study are identical to those used in the NUREG-1150 study. Sample Problem A, which is documented in the MACCS2 User's manual (Channin et al., 1990), corresponds to one of the NUREG-1150 calculations that were performed for Surry. For brevity, only the parameters that differ from those used in Sample Problem A are discussed here. While spatial grid elements are set identical to the one used in the NUREG-1150 study for Surry plant, dose factors are updated based on Federal Guidance Report (No.13) by the US Environmental Protections Agency (EPA, 1999).

Calculations performed for the NUREG-1150 study simplistically considered a single aerosol bin with a deposition velocity of 1 cm/s. Currently, MELCOR calculates aerosol size distributions using a 10-bin representation. Here, MELCOR size-distribution data are used to better quantify aerosol deposition rates. Table 5.6 shows wet and dry deposition velocities applied in this study (U.S.NRC, 2010).

Table 5.6:   Aerosol Deposition Velocity

# Bin	Dry and Wet Deposition Velocity (m/s)
1	8.58E-04
2	7.88E-04
3	1.03E-03
4	1.74E-03
5	3.40E-03
6	6.96E-03
7	1.34E-02
8	2.20E-02
9	2.73E-02
10	2.73E-02

5.2.6 MACCS2 Site Specific Input Parameters

The site specific parameters typically involve source term (also accident specific), physical dimensions of the plant building complex, and meteorological and demographic data specific to the plant region.

The source term generated by MELCOR was evaluated using MELMACCS (McFadden et al., 2005), a tool developed by Sandia to generate MACCS2 input from MELCOR calculations. MELMACCS reads a MELCOR output file containing initial and released mass of radioactive materials. MELMACCS requires a number of user specified inputs such as fission product classes to be analyzed, assumed burnup used to calculate the fission product inventory, and the initial plume dimensions.

The analysis includes the standard set of fission product classes (i.e., Xe, Cs, Ba, I, Te, Ru, Mo, Ce, and La), and the initial radionuclide inventory was calculated by MELMACCS with an assumed burnup of 49 Mwd/kg at mid cycle. When specifying plume segments, releases were broken up into two-hour plume segments. For any non-significant release, such as those with less than 1 percent of the release fractions, they are broken up into longer time intervals.

Other site specific input parameters relate to the dimensions of the building from which the release occurs. These are the 1) the building height, 2) the initial value of cross-wind dispersion (σ_y), and 3) the initial value of vertical dispersion (σ_z). The building height used in this study is 50 m. Even though MELCOR uses a lower release height (16.85 m), the MACCS analyses use a conservative release height at the top of the containment dome (50 m). The initial values for σ_y and σ_z used here are 9.302 and 23.26, respectively. These values are the initial plume standard deviations for the plume dimensions based on the building dimensions.

The meteorological data involve one year of hourly windspeed, atmospheric stability, and rainfall readings recorded at the site. The population data is constructed from the year 2000 census data, then scaled by a factor of 1.0533 to adjust for growth between the years 2000 and 2005. Details on the meteorological and population data surrounding the site are discussed in Bixler et al. (2003).

Finally, the analyses include more realistic emergency response scenarios accounting for site specific evacuation time estimates as described in NUREG-1935 (U.S.NRC, 2010) . These scenarios resulted from an assessment of the variation in emergency response for the site within the emergency planning zone (EPZ; within 16.1 km, or 10 miles, of the plant) and also included evacuation and sheltering of population groups beyond the EPZ to a distance of 32.2 km (20 miles) from the site. Each emergency scenario involves a population subgroup, referred to as a cohort, that moves differently from other population groups to account for members of the public who evacuate early, those who evacuate late, those who do not evacuate, or those who decide to evacuate from areas not under an evacuation order (defined as shadow evacuation). Six cohorts were established as described below.

Cohort 1: 0 to 10 mile Schools. This cohort includes school populations within the EPZ.

Cohort 2: 0 to 10 Special Needs and 0 to 10 Tail. This group includes two distinct population subgroups that were combined because of the small population fraction of each individual group. The Special Needs population includes residents of hospitals, nursing homes, assisted living communities and prisons. The Tail is defined as the last 10% of the public to evacuate. The 0 to 10 Special Needs and 0 to 10 Tail groups were combined because they mobilize in a similar manner and would both exit the area last. The Special Needs cohort is assumed to have a greater shielding factor during the sheltering phase prior to evacuating due to the expectation that they would likely be indoors within a facility more protective than an average home.

Cohort 3: **0 to 10 Public and 10 to 20 Shadow.** This cohort includes the public residing within the EPZ and the shadow evacuation from the 10 to 20 mile area beyond the EPZ. The Shadow population from 10 to 20 miles is assumed to begin movement at the same time as the 0 to 10 Public.

Cohort 4: 10 to 20 Public. This cohort was established to assess sheltering of residents beyond the EPZ and to support sensitivity analyses. Cohort 4 is the largest population group, and consists of the public in the region from 10 to 20 miles beyond the EPZ.

Cohort 5: 10 to 20 Special Needs, 10 to 20 Tail. This cohort was also established to assess sheltering of this population group and to support sensitivity analyses and is the last population group to begin to evacuate.

Cohort 6: **Non-evacuating public.** This cohort group represents the portion of the public who refuse to evacuate, assumed for this project as 0.5% of the total population.

Each cohort was then assigned with population fractions within 20 miles (32.2 km) (though some cohorts only include populations from 0 to 10 miles or 10 to 20 miles) as described in the following table.

Table 5.7: Cohort Population Fractions

Cohort	Description	Population Fraction
1	0 to 10 mile schools	0.061
2	0 to 10 special needs and 0 to 10 tail	0.021
3	0 to 10 public and 10 to 20 shadow	0.369
4	10 to 20 public	0.467
5	10 to 20 special needs and 10 to 20 tail	0.077
6	0 to 20 non-evacuating public	0.005

As mentioned previously, emergency response scenarios are characterized by the timing of actions taken by onsite and offsite response organizations to protect the general public by evacuation and sheltering. This is represented in evacuation time estimates (ETE). ETE were developed based on the Surry ETE report and are further described in NUREG-1935 (U.S.NRC, 2010)

The spatial movement of cohorts is then simulated using WinMACCS (McFadden et al, 2007) which allows for the movement by grid element while accounting for the speed and direction of the evacuating cohorts. The roadway network was also modeled in order to simulate evacuations and traffic movement. Detailed discussions on the development of the roadway network is also provided in NUREG-1935 (U.S.NRC, 2010)

5.2.7 MACCS2 Station Blackout Severe Accident Analysis Results

The following section presents MACCS2 consequence results for postulated Surry Station Blackout Accident scenarios. While MACCS2 consequence model calculates a large number of different consequence measures including economic consequence measures, only the results of the latent cancer fatality risk (within a 10 miles, or 16.1 km, radius from the plant) is used. This consequence measure is defined as the mean probability of dying from cancer due to the accident for an individual within 10 miles, or 16.1 km, of the plant (Breeding et al. 1990). Additionally, in order to assess the significance of the increase in health effects compared to the NRC Quantative Health Objectives (QHOs), CDF values for both STSBO and LTSBO are multiplied by the results. The corresponding CDF values for STSBO and LTSBO sequences in NUREG-1150 are 5.4E-06 and 2.2E-05, respectively.

It should be noted that the results presented here are strictly based on linear-no-threshold hypothesis (LNTH), where an assumption is made that the relationship between exposure to ionizing radiation and human cancer risk is linear. Since the focus of this study is placed on relative health consequence outcomes resulting from different levels of containment degradation, no other dose truncation is considered for low dose rates. Additionally, the selection of LNTH assumption over the dose truncation does not represent the view of authors on how one should treat the health effects for low dose rates. The LNTH model is simply chosen for shorter calculation times and simplicity in setting up the problem and easier comparison of the results between different runs.

A total of 15 different corrosion conditions (see Table 5.3) were evaluated for both STSBO and LTSBO scenarios at 96 hours. Results are presented in Table 5.8 and Table 5.9. Using the site meteorological data, MACCS2 generated output in terms of probability distributions of the consequence estimates resulting from the statistical variability in the seasonal and meteorological conditions during the accident. Values reported here represent arithmetic averages over the annual weather data used in the analysis.

Table 5.8:   STSBO Health Effects

Corrosion Level (see Table 5.3 for case descriptions)	Individual Lat. Can. Fat. Risk, 10 mi (16.1 km)	STSBO CDF Multiplied
N1	3.90E-04	2.11E-09
N2	4.36E-04	2.35E-09
N3	4.40E-04	2.38E-09
N4	4.80E-04	2.59E-09
N5	5.02E-04	2.71E-09
LB1	4.79E-04	2.59E-09
LB2	4.87E-04	2.63E-09
LB3	4.77E-04	2.58E-09
LB4	4.88E-04	2.64E-09
LB5	5.04E-04	2.72E-09
UB1	2.11E-04	1.14E-09
UB2	2.11E-04	1.14E-09
UB3	2.00E-04	1.08E-09
UB4	2.99E-04	1.61E-09
UB5	3.19E-04	1.72E-09

Table 5.9: LTSBO Health Effects

Corrosion Level (see Table 5.3 for case descriptions)	Individual Lat. Can. Fat. Risk, 10 mi (16.1 km)	LTSBO CDF Multiplied
N1	7.64E-05	1.68E-09
N2	7.90E-05	1.74E-09
N3	8.00E-05	1.76E-09
N4	9.43E-05	2.07E-09
N5	8.90E-05	1.96E-09
LB1	1.06E-04	2.33E-09
LB2	1.07E-04	2.35E-09
LB3	1.08E-04	2.38E-09
LB4	1.26E-04	2.77E-09
LB5	1.53E-04	3.37E-09
UB1	3.92E-05	8.62E-10
UB2	3.86E-05	8.49E-10
UB3	3.71E-05	8.16E-10
UB4	4.54E-05	9.99E-10
UB5	4.82E-05	1.06E-09

While the results produced with the site specific meteorological data addressed uncertainty in the weather, it was difficult to draw conclusions from this set of results due to abnormalities observed in some cases. For example, there were no apparent differences in the arithmetic average results between cases UB1 and UB2 for STSBO although it was expected to be somewhat different based on the levels of corrosion applied in those cases. However, further examination of the results showed that differences were only found in some quartiles. Because of this, it was decided to re-run the cases using only a single weather trial so that the results produced could be compared in a more deterministic fashion.

Results obtained using a single weather trial are presented in Table 5.10 and Table 5.11. Note that the single weather trial cases yield consequence results on the same order as the averaged consequence results over all of the weather cases. However, the single weather trial case risks are given that that weather trial occurs. The specific weather trial used assumed a 2 m/s wind speed and a Class F stability. In the averaged weather results, the wind speed varies from case to case between different intervals from 0 to 1 m/s, to 7 m/s and greater. The stability classes vary from A to F to determine the plume meander factor ("A" is very unstable, "F" is moderately stable).

Table 5.10:　STSBO Health Effects (single weather trial)

Corrosion Level (see Table 5.3 for case descriptions)	Individual Lat. Can. Fat. Risk, 10 mi (16.1 km)	STSBO CDF Multiplied
N1	1.55E-04	8.37E-10
N2	1.92E-04	1.04E-09
N3	1.92E-04	1.04E-09
N4	2.64E-04	1.43E-09
N5	2.50E-04	1.35E-09
LB1	2.14E-04	1.16E-09
LB2	2.34E-04	1.26E-09
LB3	2.16E-04	1.17E-09
LB4	2.28E-04	1.23E-09
LB5	2.59E-04	1.40E-09
UB1	1.47E-04	7.94E-10
UB2	1.48E-04	7.99E-10
UB3	1.41E-04	7.61E-10
UB4	1.92E-04	1.04E-09
UB5	2.11E-04	1.14E-09

Table 5.11:　LTSBO Health Effects (single weather trial)

Corrosion Level (see Table 5.3 for case descriptions)	Individual Lat. Can. Fat. Risk, 10 mi (16.1 km)	LTSBO CDF Multiplied
N1	5.74E-05	1.26E-09
N2	5.93E-05	1.30E-09
N3	5.93E-05	1.30E-09
N4	7.01E-05	1.54E-09
N5	6.62E-05	1.46E-09
LB1	7.84E-05	1.72E-09
LB2	8.04E-05	1.77E-09
LB3	8.03E-05	1.77E-09
LB4	9.38E-05	2.06E-09
LB5	1.10E-04	2.42E-09
UB1	3.09E-05	6.80E-10
UB2	3.09E-05	6.80E-10
UB3	2.98E-05	6.56E-10
UB4	3.58E-05	7.88E-10
UB5	3.75E-05	8.25E-10

Given the NRC QHO of 2×10^{-6} fatalities per year for latent cancer fatalities (NRC, 1989), the results obtained here do not indicate a risk significance in the increases observed between different levels of corrosion. In some cases, no changes were seen in the consequence results of different levels of degradation. In one case, a decrease in health consequence was observed with more severe degradation. Because of this, no direct conclusion was drawn with regards to the NRC QHOs, other than to the extent that the increase in health consequences is not significant compared to the QHOs for the cases examined.

It was anticipated that the latent cancer risks as well as other health consequences would be proportional to the extent of the corrosion. The logic behind this being that more corrosion would yield larger containment failure areas, result in higher source term releases to the environment, and hence higher consequences. Four parameters were evaluated with respect to consequences for the single weather trial in Table 5.10 and Table 5.11 (specifically with respect to the 10-mile, or 16.1 km, individual latent cancer fatality risk) to determine the validity of this supposition.

- total mass of Cesium released to the environment[2] (Figure 5.32 and Figure 5.33)
- peak containment pressure (Figure 5.34 and Figure 5.35)
- maximum leak area (Figure 5.36 and Figure 5.37)
- initial pressure at which containment failure occurs (Figure 5.38 and Figure 5.39)

Inspection of the results finds the following general trends:

- the consequence is proportional to the source term release to the environment
- the more corrosion area in most, but not all, cases resulted in higher consequences
- higher maximum containment failure area in most, but not all, cases resulted in higher consequences
- lower initial containment failure pressure in most, but not all, cases resulted in higher consequences

A detailed examination of cases UB1 and UB3 is made to ascertain the causes for case UB3, while having added a corrosion area (leaks at a lower pressure), not having a higher consequence in comparison to case UB1. Examination of the containment pressure (Figure 5.40) and containment failure area (Figure 5.41) for cases UB1 and UB3 finds that for case UB3, the initial containment failure occurs earlier (~35.5 hr vs. ~38.6 hr) than for case UB1. The early venting of the containment that occurs as a result causes case UB3 to have a lower containment pressure post-failure in comparison to case UB1. This in turn causes the failure area (crack area) for case UB3 to be lower than that for UB1 (once failure occurs in UB1). The combination of a smaller failure area and lower containment pressure in case UB3 compared to case UB1 results in less source term being released to the environment (as confirmed by comparing the two cases in Figure 5.32) and hence a lower consequence. This comparison illustrates that issues other than the extent of corrosion (in this particular case the initiation time of the failure) impact the consequence results.

---

[2] An evaluation of the MACCS2 results found that the Cesium mass released to the environment was the dominant contributor to the risk results.

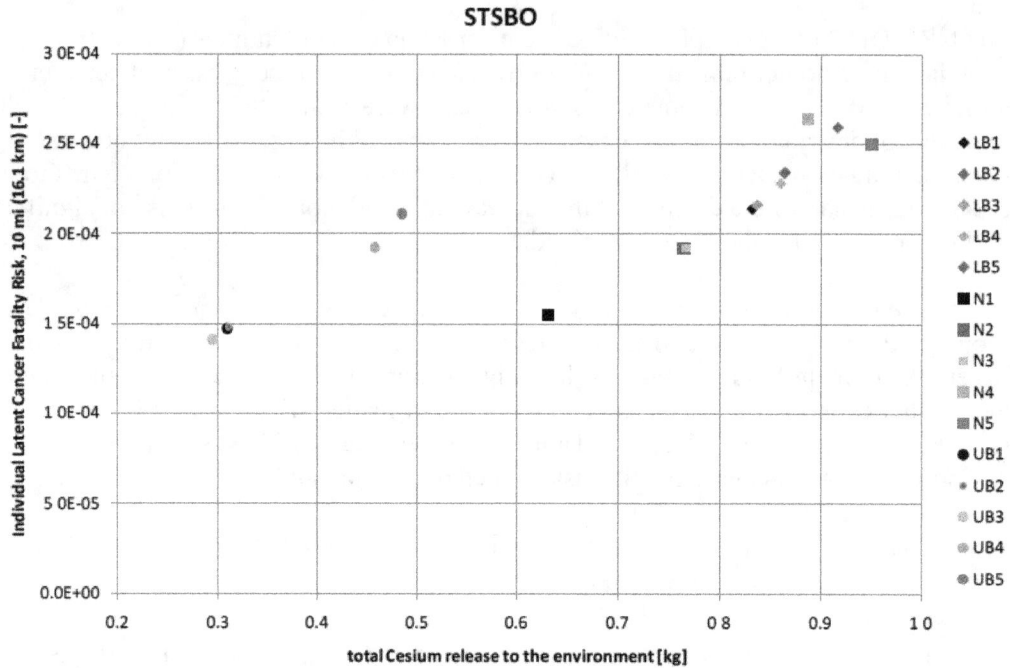

Figure 5.32: Individual Latent Cancer Facility Risk (10 Miles) as a Function of Total Cesium Release to the Environment for STSBO Cases

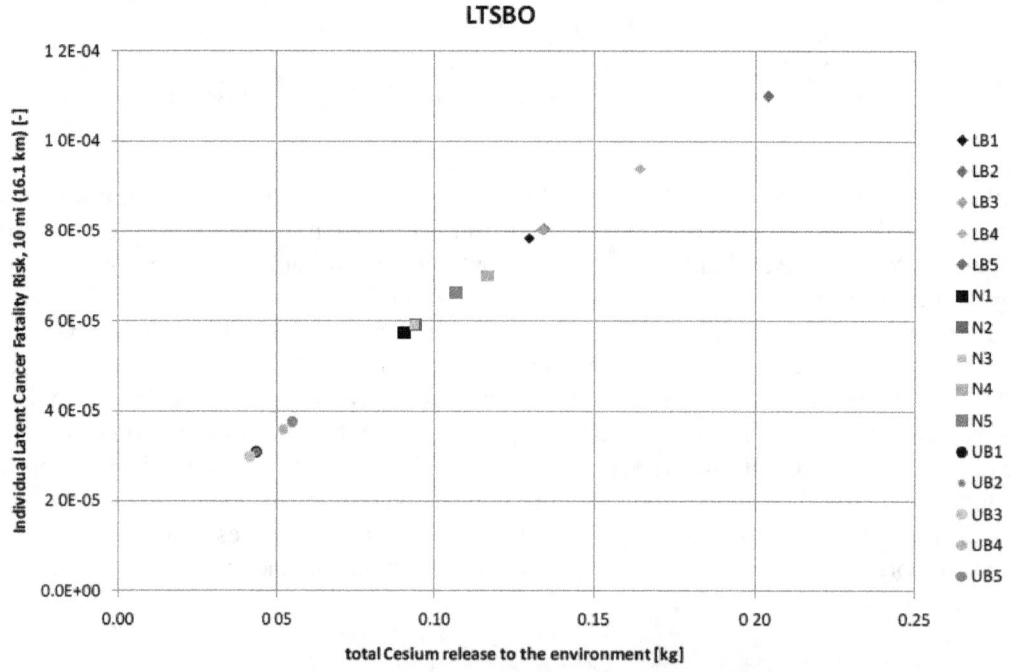

Figure 5.33: Individual Latent Cancer Facility Risk (10 Miles) as a Function of Total Cesium Release to the Environment for LTSBO Cases

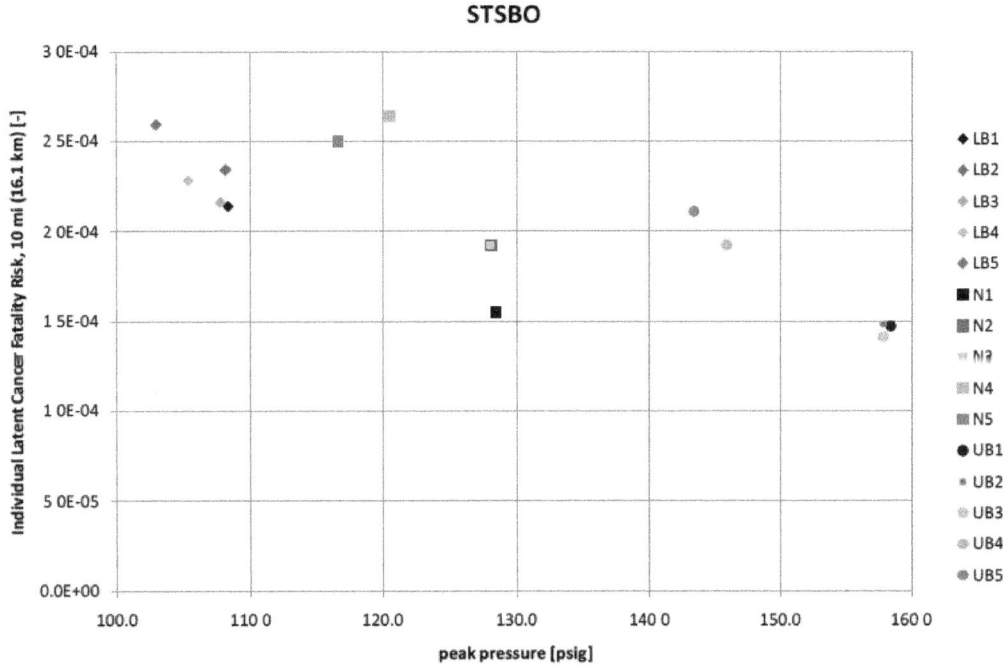

Figure 5.34: Individual Latent Cancer Facility Risk (10 Miles) as a Function of Peak Containment Pressure for STSBO Cases

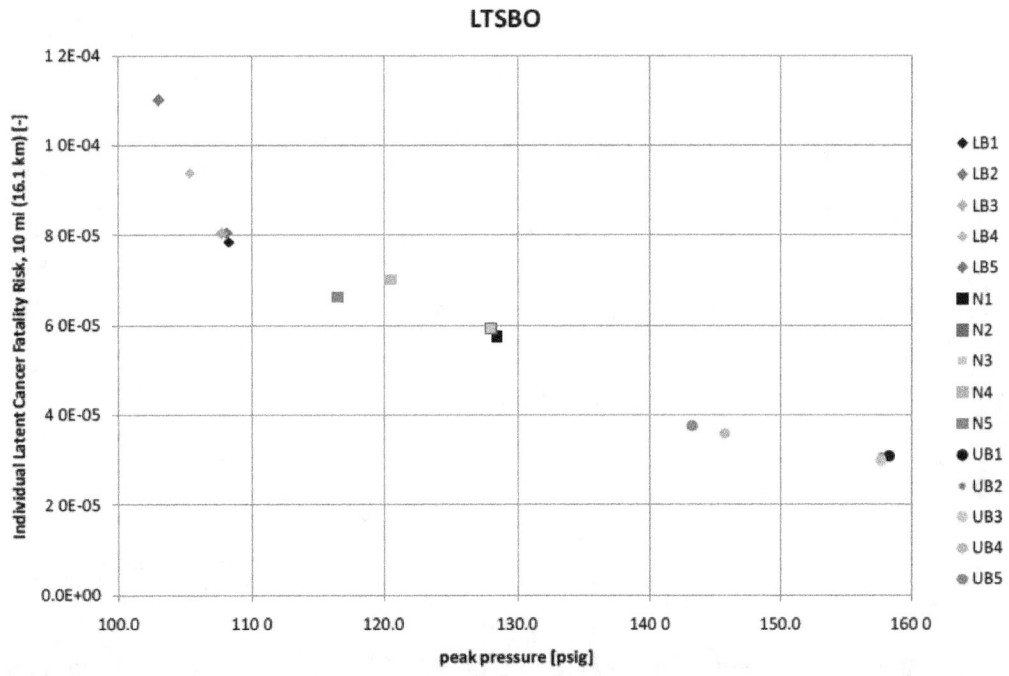

Figure 5.35: Individual Latent Cancer Facility Risk (10 Miles) as a Function of Peak Containment Pressure for LTSBO Cases

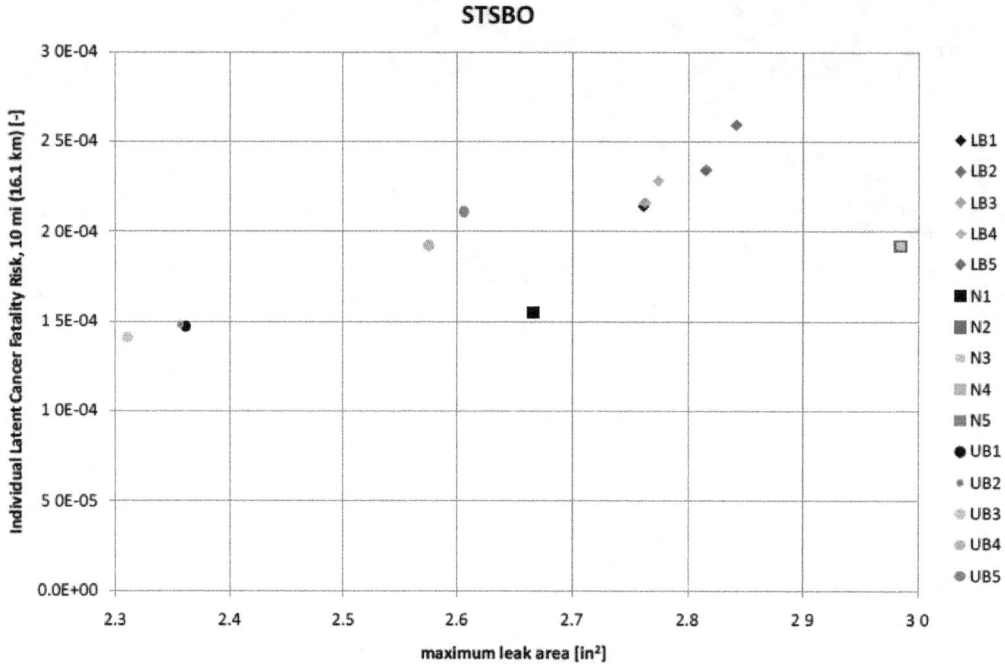

Figure 5.36: Individual Latent Cancer Facility Risk (10 Miles) as a Function of Maximum Containment Leakage Area for STSBO Cases

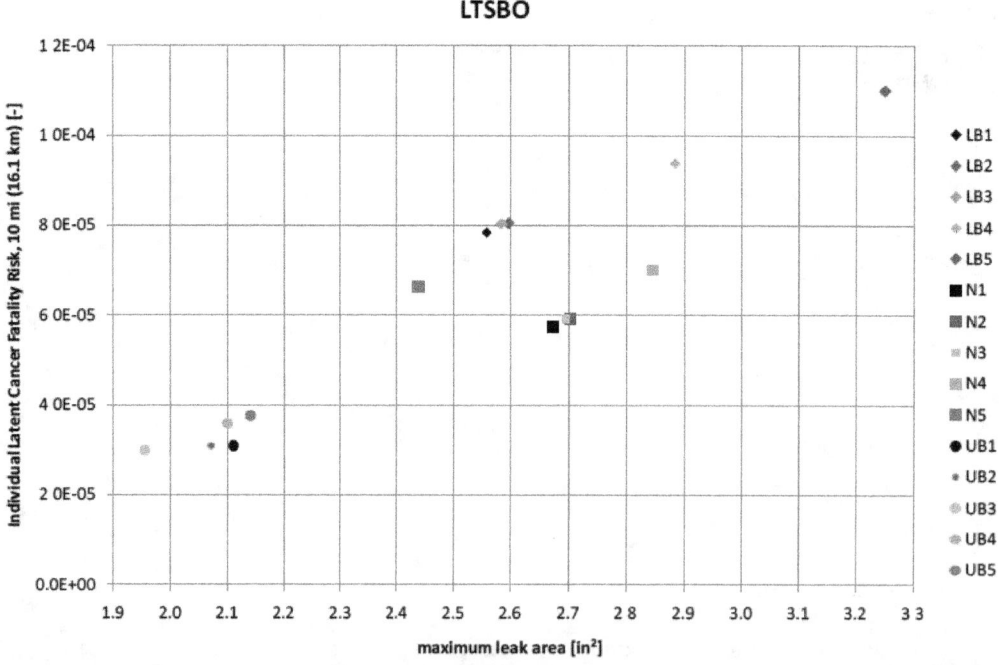

Figure 5.37: Individual Latent Cancer Facility Risk (10 Miles) as a Function of Maximum Containment Leakage Area for LTSBO Cases

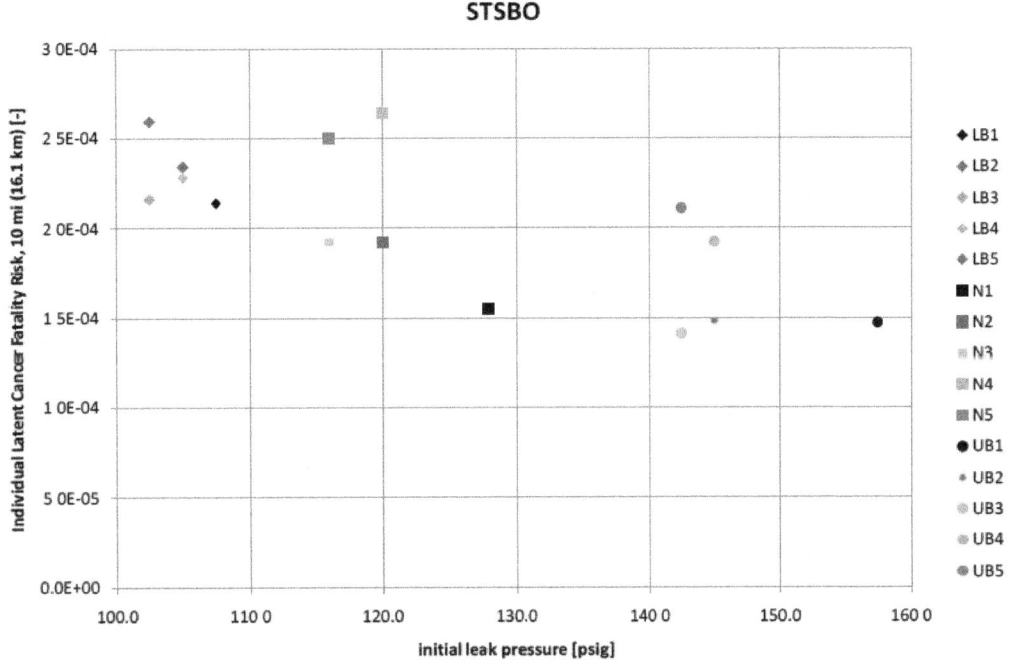

Figure 5.38: Individual Latent Cancer Facility Risk (10 Miles) as a Function of the Pressure at which Containment Failure Initiates for STSBO Cases

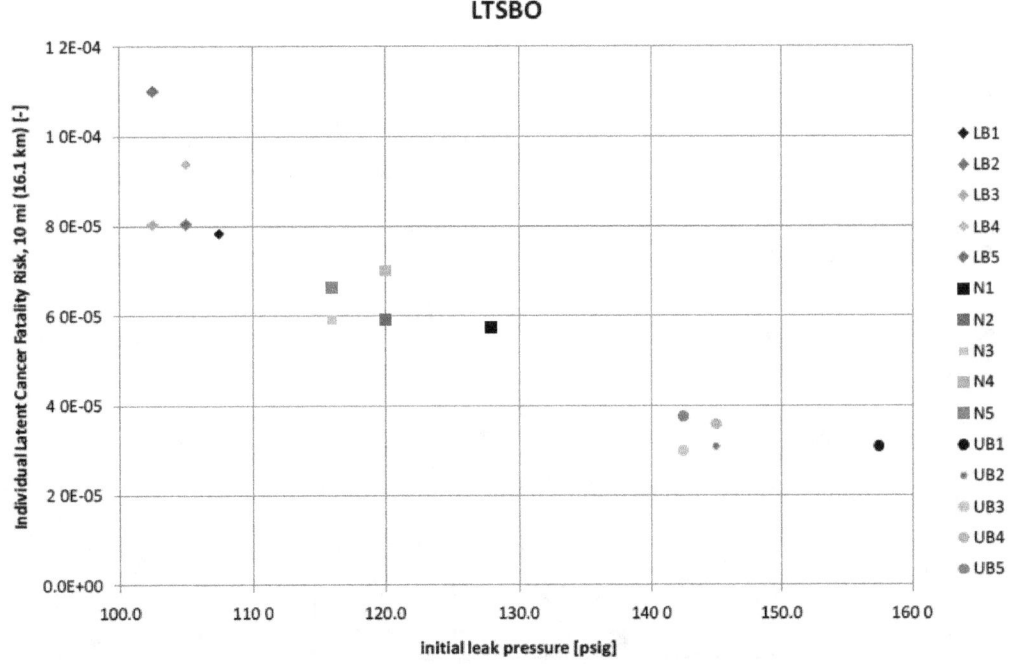

Figure 5.39: Individual Latent Cancer Facility Risk (10 Miles) as a Function of the Pressure at which Containment Failure Initiates for STSBO Cases

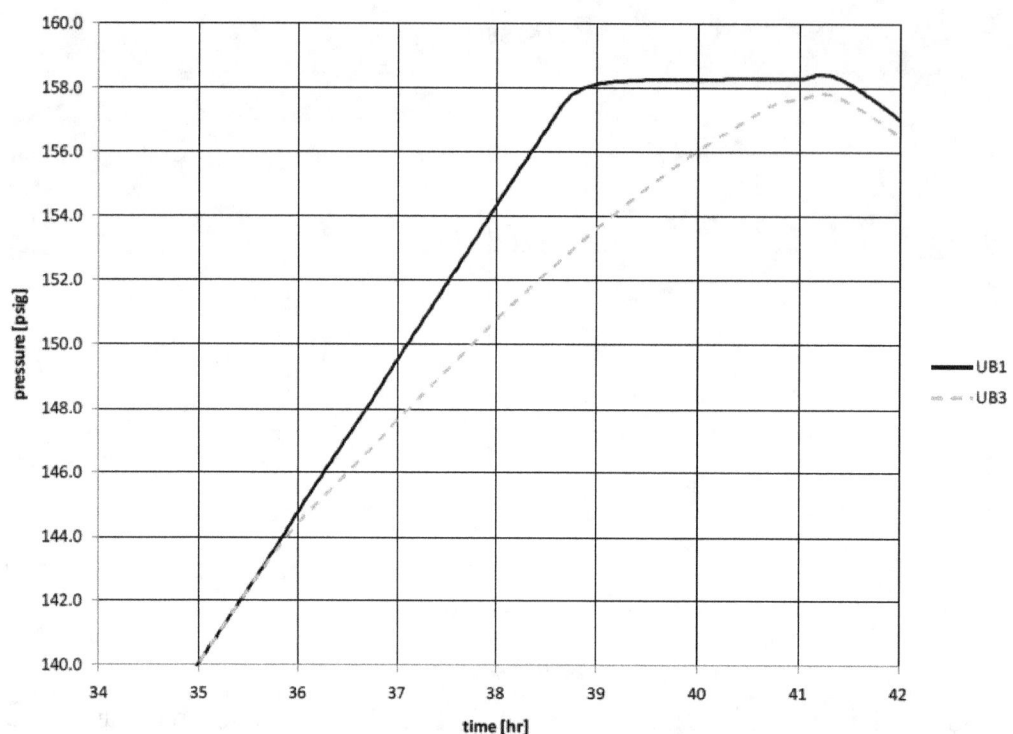

Figure 5.40: Containment Pressure for STSBO Cases UB1 and UB3

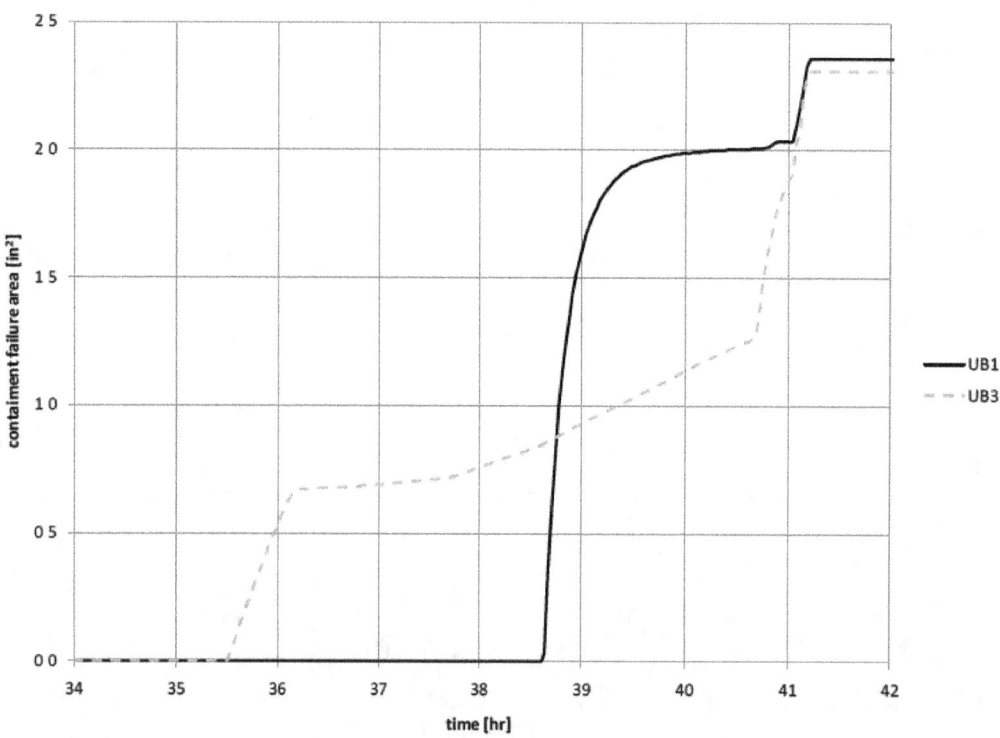

Figure 5.41: Containment Failure Area for STSBO Cases UB1 and UB3

64

5.3 PRA Analysis

The PRA analyses performed in the study by Spencer et al. (2006) are reported here but with the addition of three cases: 10 locations of 50% corrosion at the basemat, 10 locations of 50% corrosion at the midheight, and 10 locations of 65% corrosion at the midheight. The procedures outlined in Section 4.2 were used to develop the fragility curves and conditional probability of failure curves illustrated in Section 5.1.4. These curves were used in the NUREG-1150 PRA models to determine the effects of these three new cases. The basemat corrosion cases were included in the PRA analysis since some differences from the no corrosion case result. This was not the case in the MELCOR/MACCS analyses, and therefore, none of the basemat corrosion cases were including in those results. The reason for this lies in the fact that the MELCOR/MACCS analyses are deterministic, so for the cases examined, there was no noticeable difference in the computed leak areas for the basemat cases. However, when considering all 30 of the analyses used to develop the fragility and conditional probability of failure curves, a small difference was seen between the no corrosion case and the basemat corrosion cases.

Table 5.12 shows the CDF values from NUREG-1150. Because the containment response does not affect the CDF for this plant, the results are the same for all of the cases studied. These CDF values are the total for all accident scenarios considered in NUREG-1150.

Table 5.12:   Core Damage Frequency (yr^{-1})

Case	Mean	5%	50%	95%
All	4.1E-5	6.8E-6	2.3E-5	1.3E-4

Table 5.13 shows the LERF results, and Table 5.14 shows the probability of large early containment failure conditional on core damage. LERF is the product of the CDF and the probability of large early containment failure conditional on core damage. In Table 5.13 and Table 5.14, the "Un-damaged (NUREG-1150 Reference)" LERF values are also included. These values were those computed in the original NUREG-1150 study. That study used expert elicitations to develop the containment fragility curves. The current study employs the finite element modeling methods described in this report. As would be expected, there is a slight difference between the two LERF values. In addition, the Mean, 5%, 50%, and 95% values are reported in these tables. This variation is due solely to the uncertainties introduced in the original NUREG-1150 PRA study.

Table 5.13:   Large Early Release Frequency (yr^{-1})

Case	Mean	5%	50%	95%
Un-damaged (NUREG-1150 Reference)	1.10E-7	0	3.61E-10	1.54E-6
No Corrosion (Current Study)	1.00E-7	0	1.65E-10	1.12E-6
50% Corrosion near basemat	1.00E-7	0	1.65E-10	1.12E-6
50% Corrosion at midheight	1.00E-7	0	1.65E-10	1.12E-6
65% Corrosion near basemat	1.00E-7	0	1.65E-10	1.12E-6
65% Corrosion at midheight	1.00E-7	0	1.65E-10	1.12E-6
10 Regions 50% Corrosion near basemat	1.00E-7	0	1.65E-10	1.12E-6
10 Regions 50% Corrosion at midheight	1.34E-7	0	2.03E-10	1.55E-6
10 Regions 65% Corrosion at midheight	1.44E-7	0	2.03E-10	1.70E-6

Table 5.14:   Probability of Large Early Containment Failure Conditional on Core Damage

Case	Mean	5%	50%	95%
Un-damaged (NUREG-1150 Reference)	0.0027	0	1.57E-5	0.019
No Corrosion (Current Study)	0.0024	0	7.19E-6	0.0086
50% Corrosion near basemat	0.0024	0	7.19E-6	0.0086
50% Corrosion at midheight	0.0024	0	7.19E-6	0.0086
65% Corrosion near basemat	0.0024	0	7.19E-6	0.0086
65% Corrosion at midheight	0.0024	0	7.19E-6	0.0086
10 Regions 50% Corrosion near basemat	0.0024	0	7.19E-6	0.0086
10 Regions 50% Corrosion at midheight	0.0033	0	8.84E-6	0.0119
10 Regions 65% Corrosion at midheight	0.0035	0	8.84E-6	0.0130

Table 5.15:   Change in the Large Early Release Frequency (yr^{-1}) due to Degradation

Case	Mean	5%	50%	95%
50% Corrosion near basemat	0	0	0	0
50% Corrosion at midheight	0	0	0	0
65% Corrosion near basemat	0	0	0	0
65% Corrosion at midheight	0	0	0	0
10 Regions 50% Corrosion near basemat	0	0	0	0
10 Regions 50% Corrosion at midheight	3.34E-8	0	3.81E-11	4.27E-7
10 Regions 65% Corrosion at midheight	4.33E-8	0	3.81E-11	5.74E-7

The key risk criteria for acceptance of a proposed plant change in Regulatory Guide 1.174 (USNRC, 2001) are the baseline values of CDF and LERF, and the changes in those values due to the proposed modification, denoted as ΔCDF and ΔLERF. The acceptance criteria must be met for both the pairs of ΔCDF and CDF and of ΔLERF and LERF. Because CDF is assumed to be unaffected by the containment properties in this particular plant, ΔCDF is 0.

For mean LERF values, there is no change in LERF relative to the undamaged containment in the case where corrosion exists near the basemat. In addition, the single cases of corrosion at the midheight also do not cause any increase in LERF. However, the two cases with 10 regions of corrosion at the midheight did cause a measurable increase in LERF as shown in Table 5.15. These increases in LERF are small and still fall into Region III for the LERF acceptance criteria in Figure 4.1. Therefore, the corrosion cases investigated in this study would be permissible modifications from a risk standpoint by Regulatory Guide 1.174. However, the addition of the new cases with more widespread degradation might show a greater effect.

Even so, the results presented above can potentially be somewhat misleading. One would intuitively expect that at least a small increase in the risk measure would be observed in the cases with any level of containment degradation. As expected, these increases were manifested in at least some of the deterministic MELCOR/MACCS analyses. This lack of increase in the risk measures for the PRA analyses presented in Table 5.13 and Table 5.14 can be traced to the definition of LERF used in the process of binning the risk analysis results. LERF is, by definition, the frequency of large, early containment failures resulting from accidents stemming from a certain set of initiating events. The question of whether the failure is large is determined by the mode of failure, which is computed based on the combination of the overall fragility curve

66

and the conditional probabilities of leak, rupture, and catastrophic rupture. Failure occurrences are grouped into LERF only if the containment fails in a rupture or catastrophic rupture mode.

As examined in the Spencer study, the Small Early Release Frequency (SERF) and the Total Early Release Frequency (TERF) were also computed. TERF is defined as the sum of LERF and SERF. The PRA analyses bin the leak failures into SERF which is useful in examining what information is left out when analyzing the LERF data alone. Table 5.16 and Table 5.17 summarize the SERF and TERF mean values for each of the cases examined as well as the change in each of the degraded cases from the no corrosion case. Due to the liner corrosion examined in this study leading to leak failures, the change in SERF from the no corrosion case is more pronounced than for LERF. For the cases where ΔLERF equal 0, ΔSERF and ΔTERF will be equal. Most interesting is that the largest increase in SERF is for the 65% Corrosion at midheight case. However, the TERF for the 65% Corrosion at the midheight and the same case with 10 regions of corrosion are the same. This shows that the failures for the single corrosion case were all under the rupture limit (e.g. leaks) and were all binned into SERF. The case with 10 regions of corrosion had failures split between leak and rupture causing increases in both SERF and LERF. As discussed in the Spencer study, there is no regulatory guidance or limits on smaller leak failures binned into SERF. The MELCOR/MACCS analyses showed that these small earlier leaks do potentially cause an increase in consequences.

Table 5.16:   Small Early Release Frequency (yr^{-1}) and the Change in SERF

Case	Mean SERF	ΔSERF
No Corrosion	4.36E-8	N/A
50% Corrosion near basemat	4.62E-8	2.58E-9
50% Corrosion at midheight	1.14E-7	7.01E-8
65% Corrosion near basemat	4.62E-8	2.58E-9
65% Corrosion at midheight	1.45E-7	1.02E-7
10 Regions 50% Corrosion near basemat	4.62E-8	2.58E-9
10 Regions 50% Corrosion at midheight	8.08E-8	3.73E-8
10 Regions 65% Corrosion at midheight	1.02E-7	5.83E-8

Table 5.17:   Total Early Release Frequency (yr^{-1}) and the Change in TERF

Case	Mean TERF	ΔTERF
No Corrosion	1.44E-7	N/A
50% Corrosion near basemat	1.46E-7	2.58E-9
50% Corrosion at midheight	2.14E-7	7.01E-8
65% Corrosion near basemat	1.46E-7	2.58E-9
65% Corrosion at midheight	2.45E-7	1.02E-7
10 Regions 50% Corrosion near basemat	1.46E-7	2.58E-9
10 Regions 50% Corrosion at midheight	2.14E-8	7.06E-8
10 Regions 65% Corrosion at midheight	2.45E-7	1.02E-7

The example plant examined here serves as a good case study which indicates that one should be very cautious about simply applying the ΔLERF criterion to evaluate the significance of containment degradation. It is important to closely examine how degradation can increase the likelihood of one failure mode while decreasing the likelihood of another, and evaluate the consequences of these different modes. It cannot be stressed enough that the cases of degradation

67

examined here are extremely limited in scope. There are currently no means of predicting degradation prior to its discovery. Since the location and specific nature of any degradation could dramatically affect the containment capacity and risk, a detailed analysis of that scenario would be required. There are also a number of assumptions that would require additional consideration when performing a site specific analysis. The specific results reported in this study should not be applied to making specific regulatory decisions on existing plants.

6. SUMMARY & CONCLUSIONS

This study has examined the effects of local liner degradation on the consequences during a severe accident for a reinforced concrete containment vessel at a PWR plant. Two different methods were used to compute the consequences: a series of PRA analyses to compute LERF, and deterministic analyses to compute direct health consequences. The PRA analysis method was developed in the study by Spencer et al. (2006). In that work, a series of structural analyses were conducted to develop fragility curves for the failure probability distributions for various cases of degradation. The Spencer study examined four different containments, but the current study only includes the reinforced concrete containment. The results of that study showed that local liner corrosion contributed mainly to the small early release frequency due to the binning process used in the PRA modeling. The goal of that work was to show if LERF could be used as a metric with Regulatory Guide 1.174 to assess containment degradation. The Spencer study showed that this may work when applied to cases of degradation that affect the rupture pressures (e.g. LERF contributors), but may miss potential consequences due to early, small leaks. The goal of the current study was to explore alternate metrics beyond LERF. In order to examine other metrics, a series of deterministic analyses were performed. These analyses were conducted using the Sandia-developed MELCOR and MACCS codes to perform the accident progression simulations and the resulting consequence calculations. Both the long- and short-term station black-out (LTSBO and STSBO) accident scenarios were examined. Each scenario was examined for 5 different containment conditions: no corrosion, one small patch of 50% corrosion at the midheight, one small patch of 65% corrosion at the midheight, ten small patches of 50% corrosion at the midheight, and ten small patches of 65% corrosion at the midheight. The cases of near-basemat corrosion studied for the PRA analyses performed for the Spencer study were determined to not vary from the no corrosion case due to other failures dominating, and therefore, were not examined. For each accident scenario and corrosion condition, analyses were performed using best estimate crack area vs. pressure curves computed from the structural analysis results created during the Spencer study. In addition, upper and lower bound area vs. pressure curves were also used as input for additional MELCOR/MACCS analyses. These upper and lower bound cases were computed by enveloping the area vs. pressure curve for a given corrosion case using 30 different analyses of each given degradation case. These 30 cases were used to develop the fragility curves for the PRA analysis and account for uncertainties in the crack area vs. pressure response. A number of response metrics were examined as a function of the consequences determined in MACCS for the different corrosion cases. These include:

- total mass of Cesium released to the environment (Figure 5.32 and Figure 5.33)
- peak containment pressure (Figure 5.34 and Figure 5.35)
- maximum leak area (Figure 5.36 and Figure 5.37)
- initial pressure at which containment failure occurs (Figure 5.38 and Figure 5.39)

As summarized in the MACCS results section, inspection of the results finds the following general trends:

- consequence is proportional to the source term release to the environment
- more corrosion area in most, but not all, cases resulted in higher consequences
- higher maximum containment failure area in most, but not all, cases resulted in higher consequences

- lower initial containment failure pressure in most, but not all, cases resulted in higher consequences

These are mostly general trends since cases were observed where the timing of the containment leakage affected the consequences. Higher levels of liner corrosion and the extent of that corrosion led to leakage at lower pressures and to a larger amount of crack area at lower pressures. Though in some cases, small amounts of corrosion were shown to act as a vent. This lowered the internal pressure of the containment vessel prior to the inventory of radionuclides fully developing. Therefore, when more consequence significant material was generated within the containment, the pressure could be lower than in a case with no corrosion, or less corrosion, pushing less material out to the environment.

The PRA computations performed in the Spencer study were also reexamined with the addition of cases where more extensive corrosion exists. It was found that while the small local corrosions were not large enough to cause a change in LERF, more extensive corrosion did affect the rupture pressures and LERF did increase for some cases. However, the PRA method of examining only LERF for assessment in Regulatory Guide 1.174 does miss the smaller early releases as demonstrated when examining SERF and in the MELCOR/MACCS analyses.

The example plant examined here serves as a good case study indicating that one should be very cautious about simply applying a LERF criterion or assuming a general trend exists in more deterministic severe accident simulations when evaluating the significance of containment degradation. It is important to closely examine how degradation can change the consequence response of the containment on a case-by-case basis. Due to the more detailed information and fidelity in the different responses, the deterministic accident scenarios may be more useful when examining a specific case of degradation. It cannot be stressed enough that the cases of degradation examined here are extremely limited in scope and should not be used to generalize results. In addition, the cases of corrosion examined here are postulated and the results in this report should not be used to assess actual cases of corrosion. The patches of corrosion were very small and assumed to have reduced the liner thickness by 50% or 65%. Recent interest has focused on initial through liner corrosion in concrete containments. In these cases, a hole has reduced the liner thickness by 100% over some area. The tensile limit of the concrete will be reached at very low pressures within the containment leading to leak paths through the hole in the liner and cracked concrete. Here, the timing of the accident becomes critical in the resulting consequences. The earlier leaks in the containment would not allow the pressure to increase as much as a non-degraded liner. Therefore, the amount of high consequence material created later in the accident will not be ejected to the environment at the same rate. However, the total time of release to the environment will increase. There is no way to develop a general conclusion as to the effects of initial through holes or cracks just as there is no method to generalize any case of degradation. Most power plants in the U.S. are unique, which requires a case by case examination of the containment structural response and the examination of plausible site specific accident scenarios. There are also a number of assumptions that would require additional consideration when performing a site specific analysis. In addition, there are other accident scenarios that would require assessment for a site specific case. These include events with rapid pressurization (e.g., hydrogen burn) among others. Therefore, the specific results reported in this study should in no way be applied to making specific regulatory decisions on specific plants.

7. REFERENCES

ANATECH, 1997, "ANACAP-U User's Manual, Version 2.5" ANATECH Corp., San Diego, CA.

Ang, A.H-S. and Tang, W.H., 1975, *Probability Concepts in Engineering Planning and Design, Volume 1-Basic Principles*, Wiley.

Bixler, N.E. et al., SECPOP2000: Sector Population, Land Fraction, and Economic Estimation Program, NUREG/CR-6525, Rev. 1, SAND2003-1684P, Sandia National Laboratories, Albuquerque, NM, 2003.

Breeding, R.J., F.T. Harper, T.D. Brown, J.J. Gregory, A.C. Payne, E.D. Gorham, W. Murfin, and C.N. Amos, 1992, "Evaluation of Severe Accident Risks: Quantification of Major Input Parameters." NUREG/CR-4551, SAND86-1309, Vol. 2, Rev. 1, Part 3, Sandia National Laboratories, Albuquerque, NM.

Breeding, R.J., J.C. Helton, W.B. Murfin, and L.N. Smith, 1990, "Evaluation of Severe Accident Risks: Surry Unit 1." NUREG/CR-4551, SAND86-1309, Vol. 3, Sandia National Laboratories, Albuquerque, NM.

Castro, J.C., R.A. Dameron, R.S. Dunham, and Y.R. Rashid, 1993, "A Probabilistic Approach for Predicting Concrete Containment Leakage." EPRI Report TR-102176, T1, ANATECH Corp., San Diego, CA.

Channin, D.I., et al., MELCOR Accident Code System (MACCS), Vol. I: User's Guide, NUREG/CR-4691, SAND86-1562, Sandia National Laboratories, Albuquerque, NM, 1990.

Channin, D.I., et al., MELCOR Accident Code System (MACCS), Vol. II: Model Description, NUREG/CR-4691, SAND86-1562, Sandia National Laboratories, Albuquerque, NM, 1990.

Channin, D.I., et al., MELCOR Accident Code System (MACCS), Vol. III: Programmer's Reference Manual, NUREG/CR-4691, Sandia National Laboratories, Albuquerque, NM, 1990.

Cherry, J.L. and J.A. Smith, 2001, "Capacity of Steel and Concrete Containment Vessels with Corrosion Damage." NUREG/CR-6706, SAND2000-1735, Sandia National Laboratories, Albuquerque, NM.

Clauss, D.B., 1987, "Round-Robin Pretest Analyses of a 1:6 scale Reinforced Concrete Containment Model Subject to Static Internal Pressurization." NUREG/CR-4913, SAND87-0891, Sandia National Laboratories, Albuquerque, NM.

Dameron, R.A., Rashid, Y.R., and H.T. Tang, 1995, "Leak area and leakage prediction for probabilistic risk assessment of concrete containments under severe core conditions." *Nuclear Engineering and Design*, 156, 173-179.

Dameron, R.A., L. Zhang, Y.R. Rashid, and M.S. Vargas, 2000, "Pretest Analysis of a 1:4-Scale Prestressed Concrete Containment Vessel Model." NUREG/CR-6685, SAND2000-2093, Sandia National Laboratories, Albuquerque, NM.

Dobbe, C.A., Letter to Dr. J.E. Kelly (Sandia National Laboratories), "Transmittal of Floppy Disk Containing the MELCOR Input Decks of the Surry PWR – CAD-2-88," Idaho National Engineering Laboratory, May 17, 1988.

Eldred, M.S., A.A. Giunta, B.G. van Bloeman Waanders, S.F. Wojtkiewicz Jr., W.E. Hart, and M.P. Alleva, 2002, "DAKOTA, A Multilevel Parallel Object-Oriented Framework for Design Optimization, Parameter, Estimation, Uncertainty Quantification, and Sensitivity Analysis. Version 3.0 Users Manual." SAND2001-3796, Sandia National Laboratories, Albuquerque, NM.

Ellingwood, B. R. and J.L. Cherry, 1999, "Fragility Modeling of Aging Containment Metallic Pressure Boundaries." NUREG/CR-6631, ORNL/SUB/99-SP638V, Oak Ridge National Laboratories, Oak Ridge, TN.

Environmental Protection Agency (EPA), "Human Health Evaluation Manual, Supplemental Guidance: Standard Default Exposure Factors", OSWER Directive 9285.6-03, PB91-921314, Washington, DC., 1991.

Environmental Protection Agency (EPA), Cancer Risk Coefficients for Environmental Exposure to Radionuclides, FGR 13, EPA 402-R-99-001, Washington, DC., 1999.

Gauntt, R., et al., 2005, *MELCOR Computer Code Manuals*, NUREG/CR-6119, Vol. 1 & 2, Rev. 3, SAND2005-5713, Washington, D.C.

Hancock, J.W. and A.C. Mackenzie 1976, "On the Mechanisms of Ductile Failure in High-Strength Steels Subjected to Multi-Axial Stress States," *Journal of Mechanics and Physics of Solids*, 24, 147-169.

Hibbit, Karlsson and Sorenson (HKS), 2002, "ABAQUS/Standard User's Manual, Version 6.3," Hibbit, Karlsson and Sorenson, Inc, Pawtucket, RI.

Hessheimer, M.F., E.W. Klamerus, L.D. Lambert, G.S. Rightley, and R.A. Dameron, 2003, "Overpressurization Test of a 1:4-Scale Prestressed Concrete Containment Vessel Model." NUREG/CR-6810, SAND2003-0840P, Sandia National Laboratories, Albuquerque, NM.

Horschel, D.S., 1992, "Experimental Results From Pressure Testing a 1:6-Scale Nuclear Power Plant Containment." NUREG/CR-5121, SAND88-0906, Sandia National Laboratories, Albuquerque, NM.

McFadden, K.L., N.E. Bixler, and R.O. Gauntt, *MELMACCS System Documentation (MELCOR to MACCS2 interface definition)*, Sandia National Laboratories, Albuquerque, NM. 2005.

McFadden, K.L. et al, WinMACCS, a MACCS2 Interface for Calculating Health and Economic Consequences from Accidental Release of Radioactive Materials into the Atmosphere, User's Guide and Reference Manual, WinMACCS Version 3, Sandia National Laboratories, Albuquerque, NM, July 2007.

McKay, M.D., R.J. Beckman, and W.J. Conover, 1979, "A Comparison of Three Methods for Selecting Values of Input Variables in the Analysis of Output from a Computer Code." *Technometrics*, 21(2),239-245.

Mirza, S.A., M. Hatzinikolas, and J.G. MacGregor, 1979, "Statistical Descriptions of Strength of Concrete." *Journal of the Structural Division*, ASCE, 105 (6),1021-1037.

Smith, J.A., 2001, "Capacity of Prestressed Concrete Containment Vessels with Prestressing Loss." SAND2001-1762, Sandia National Laboratories, Albuquerque, NM.

Spencer, B.W., J.P. Petti, and D.M. Kunsman, 2006, "Risk-Informed Assessment of Degraded Containment Vessels," NUREG/CR-6920, SAND2006-3772P, Sandia National Laboratories, Albuquerque, NM.

Systems Applications, Inc., "Survey of Plume Models for Atmospheric Application", EPRI EA-2243, Electric Power Research Institute, Palo Alto, CA, 1982.

Tang, H.T., R.A. Dameron, and Y.R. Rashid, 1995, "Probabilistic Evaluation of Concrete Containment Capacity for Beyond Design Basis Internal Pressures," *Nuclear Engineering and Design*, 157, 455-467.

United States Nuclear Regulatory Commission (NRC), 1986, "Safety Goals for the Operations of Nuclear Power Plants; Policy Statement," *Federal Register*, Vol. 51, p. 30028 (51 FR 30028), August 4, 1986.

United States Nuclear Regulatory Commission (NRC), SECY-89-102, Implementation of the Safety Goal Policy, March 30, 1989.

United States Nuclear Regulatory Commission (NRC), 1990, "Severe Accident Risks: An Assessment for Five U.S. Nuclear Power Plants", NUREG-1150.

United States Nuclear Regulatory Commission (NRC), 2012, "State-of-the-Art Reactor Consequence Analyses (SOARCA) Report – Draft Report for Comment", NUREG-1935.

United States Nuclear Regulatory Commission (NRC), 2012, "State-of-the-Art Reactor Consequence Analyses Project – Volume 2: Surry Integrated Analysis", NUREG/CR-7110, Vol. 2.

United States Nuclear Regulatory Commission (NRC), 1995, "Use of Probabilistic Risk Assessment Methods in Nuclear Activities: Final Policy Statement," *Federal Register*, Vol. 60, p. 42622, (60 FR 42622), August 16, 1995.

United States Nuclear Regulatory Commission (NRC), 2001, "An Approach for Using Probabilistic Risk Assessment in Risk-Informed Decisions on Plant-Specific Changes to the Licensing Basis." Regulatory Guide 1.174 (Draft Regulatory Guide DG-1110).

NRC FORM 335 (12-2010) NRCMD 3.7	U.S. NUCLEAR REGULATORY COMMISSION	1. REPORT NUMBER (Assigned by NRC, Add Vol., Supp., Rev., and Addendum Numbers, if any.) NUREG/CR-7149
	BIBLIOGRAPHIC DATA SHEET *(See instructions on the reverse)*	

2. TITLE AND SUBTITLE	3. DATE REPORT PUBLISHED	
Effects of Degradation on the Severe Accident Consequences for a PWR Plant with a Reinforced Concrete Containment Vessel	MONTH June	YEAR 2013
	4. FIN OR GRANT NUMBER	

5. AUTHOR(S)	6. TYPE OF REPORT
Petti, J. P., Kalinich, D. A., Jun, J., Wagner, K. C.	Final technical report
	7. PERIOD COVERED (Inclusive Dates) January 2008 to September 2012

8. PERFORMING ORGANIZATION - NAME AND ADDRESS (If NRC, provide Division, Office or Region, U. S. Nuclear Regulatory Commission, and mailing address; if contractor, provide name and mailing address.)

Sandia National Laboratories
Albuquerque, NM 87185

9. SPONSORING ORGANIZATION - NAME AND ADDRESS (If NRC, type "Same as above", if contractor, provide NRC Division, Office or Region, U. S. Nuclear Regulatory Commission, and mailing address.)

U.S. Nuclear Regulatory Commission
Office of Nuclear Regulatory Division

10. SUPPLEMENTARY NOTES

11. ABSTRACT (200 words or less)

This report documents a methodology to assess the effects of containment degradation in terms of consequences using metrics other than Large Early Release Frequency (LERF). It is the last report on a study to examine what degree of degradation a containment can have and still be left in service without repair, and not have a significant impact on risk. The goal of the study was to enhance understanding of containment safety margins for degraded conditions. The overall study examined effects of degradation in containments using deterministic and risk-informed methods. The report documents a methodology that integrates containment fragility functions for non-degraded and postulated degraded conditions with accident progression analyses and offsite consequence assessments made with the Sandia National Laboratories codes MELCOR and MACCS. Illustrative applications simulated two accident scenarios and computed changes in consequences for a PWR plant with non-degraded and degraded reinforced concrete containments. Degraded conditions considered corresponded to several postulated cases of liner corrosion involving location, depth and extent of the corrosion to compare changes in consequences for these cases. The report presents and illustrates a methodology that is valuable to inform case by case examination of containment vessel degradation effects.

12. KEY WORDS/DESCRIPTORS (List words or phrases that will assist researchers in locating the report.)	13. AVAILABILITY STATEMENT
Containment Degradation; Containment Liners; Containment Liner Corrosion; Containment Fragility; Deterministic Methods; MACCS; MELCOR; Nuclear Power Plant Containment Buildings; Pressurized Water Reactor (PWR); Reactor Consequences; Reinforced Concrete; Risk-informed Methods; Severe Accident; Structural Analysis.	unlimited
	14. SECURITY CLASSIFICATION
	(This Page) unclassified
	(This Report) unclassified
	15. NUMBER OF PAGES
	16. PRICE

UNITED STATES
NUCLEAR REGULATORY COMMISSION
WASHINGTON, DC 20555-0001

OFFICIAL BUSINESS

NUREG/CR-7149

Effects of Degradation on the Severe Accident Consequences for a PWR Plant with a Reinforced Concrete Containment Vessel

June 2013

www.ingramcontent.com/pod-product-compliance
Lightning Source LLC
Chambersburg PA
CBHW081831170526
45167CB00007B/2790